Transfer Learning for Multiagent Reinforcement Learning Systems

Synthesis Lectures on Artificial Intelligence and Machine Learning

Editors

Ronald Brachman, *Jacobs Technion–Cornell Institute at Cornell Tech*
Francesca Rossi, *IBM Research AI*
Peter Stone, *University of Texas at Austin*

Series Page

Metric Learning
Aurélien Bellet, Amaury Habrard, and Marc Sebban
2015

Graph-Based Semi-Supervised Learning
Amarnag Subramanya and Partha Pratim Talukdar
2014

Robot Learning from Human Teachers
Sonia Chernova and Andrea L. Thomaz
2014

General Game Playing
Michael Genesereth and Michael Thielscher
2014

Judgment Aggregation: A Primer
Davide Grossi and Gabriella Pigozzi
2014

An Introduction to Constraint-Based Temporal Reasoning
Roman Barták, Robert A. Morris, and K. Brent Venable
2014

Reasoning with Probabilistic and Deterministic Graphical Models: Exact Algorithms
Rina Dechter
2013

Introduction to Intelligent Systems in Traffic and Transportation
Ana L.C. Bazzan and Franziska Klügl
2013

A Concise Introduction to Models and Methods for Automated Planning
Hector Geffner and Blai Bonet
2013

Essential Principles for Autonomous Robotics
Henry Hexmoor
2013

Case-Based Reasoning: A Concise Introduction
Beatriz López
2013

Answer Set Solving in Practice
Martin Gebser, Roland Kaminski, Benjamin Kaufmann, and Torsten Schaub
2012

Transfer Learning for Multiagent Reinforcement Learning Systems
Felipe Leno da Silva and Anna Helena Reali Costa

ISBN: 978-3-031-00463-6   paperback
ISBN: 978-3-031-01591-5   ebook
ISBN: 978-3-031-00036-2   hardcover

DOI 10.1007/978-3-031-01591-5

A Publication in the Springer series
SYNTHESIS LECTURES ON ARTIFICIAL INTELLIGENCE AND MACHINE LEARNING

Lecture #49
Series Editors: Ronald Brachman, *Jacobs Technion–Cornell Institute at Cornell Tech*
                Francesca Rossi, *IBM Research AI*
                Peter Stone, *University of Texas at Austin*
Series ISSN
Print 1939-4608   Electronic 1939-4616

Transfer Learning for Multiagent Reinforcement Learning Systems

Felipe Leno da Silva
Advanced Institute for AI

Anna Helena Reali Costa
Universidade de São Paulo

SYNTHESIS LECTURES ON ARTIFICIAL INTELLIGENCE AND MACHINE LEARNING #49

ABSTRACT

Learning to solve sequential decision-making tasks is difficult. Humans take years exploring the environment essentially in a random way until they are able to reason, solve difficult tasks, and collaborate with other humans towards a common goal. Artificial Intelligent agents are like humans in this aspect. Reinforcement Learning (RL) is a well-known technique to train autonomous agents through interactions with the environment. Unfortunately, the learning process has a high sample complexity to infer an effective actuation policy, especially when multiple agents are simultaneously actuating in the environment.

However, previous knowledge can be leveraged to accelerate learning and enable solving harder tasks. In the same way humans build skills and reuse them by relating different tasks, RL agents might reuse knowledge from previously solved tasks and from the exchange of knowledge with other agents in the environment. In fact, virtually all of the most challenging tasks currently solved by RL rely on embedded knowledge reuse techniques, such as Imitation Learning, Learning from Demonstration, and Curriculum Learning.

This book surveys the literature on knowledge reuse in multiagent RL. The authors define a unifying taxonomy of state-of-the-art solutions for reusing knowledge, providing a comprehensive discussion of recent progress in the area. In this book, readers will find a comprehensive discussion of the many ways in which knowledge can be reused in multiagent sequential decision-making tasks, as well as in which scenarios each of the approaches is more efficient. The authors also provide their view of the current low-hanging fruit developments of the area, as well as the still-open big questions that could result in breakthrough developments. Finally, the book provides resources to researchers who intend to join this area or leverage those techniques, including a list of conferences, journals, and implementation tools.

This book will be useful for a wide audience; and will hopefully promote new dialogues across communities and novel developments in the area.

KEYWORDS

multiagent reinforcement learning, transfer learning, learning from demonstrations, imitation learning, multi-task learning, knowledge reuse, machine learning, artificial intelligence

Contents

Preface

Artificial Intelligence (AI) systems are becoming so pervasive and integrated into our routine that adaptive behavior is rapidly becoming a necessity rather than an innovation in applications. *Learning* arose as the paradigm that brought AI to the spotlight, thanks to its ability to solve challenging tasks with minimal modeling effort. Personal virtual assistants, AI-powered robots, recommendation systems, smart devices, and autonomous cars were all science fiction material a couple of decades (perhaps years) ago, but are now part of our reality.

Despite recent successes, AI systems are still limited in reasoning about the long-term effects of their actions. *Reinforcement Learning* (RL) focuses on sequential decision-making problems. These techniques are based on an extensive exploration of the environment for learning action effects, which means that these techniques have a high sample complexity.

We firmly believe that collective efforts by artificial agents offer the path to crack yet-unsolved application challenges. Therefore, efficient techniques for training groups of agents to solve sequential decision-making tasks might be the next major AI breakthrough. However, in addition to the well-known high sample complexity that is amplified by the presence of multiple agents, multiagent RL has to cope with additional challenges such as the nonstationarity of other agents' behavior.

With this vision in mind, we have dedicated a good part of our scientific careers working on approaches for scaling up and facilitating the training of multiagent RL systems. As it happens with humans, reusing knowledge is a very effective way to accelerate the learning process. Hence, our approach has been to reuse knowledge to learn faster and to enable learning challenging tasks previously unsolvable through direct exploration.

The content that eventually became this book started to be written down years ago, as personal notes about the field, during the development of the first author's Ph.D. However, the material started to grow and to integrate works from different sub-communities such as *Supervised Learning*, *Multiagent Systems*, *Active Learning*, and of course *Reinforcement Learning*, each of them with their own (sometimes inconsistent) terminologies, jargon, and symbols. It was clear to us that an integrating effort was needed to systematically convey the content of the area.

In this context, we published a survey on *Transfer Learning in Multiagent Reinforcement Learning* in the *Journal of Artificial Intelligence Research* (JAIR). This book was written by extending that survey in several dimensions. First, although the survey was written only a couple of years ago, the area has had many developments after that, and we took this opportunity to update our picture of the state-of-the-art in the area. Second, the book format allowed us to reformulate the manuscript in a more didactic format. We included informative illustrations and described in more detail the background needed to fully understand the proposals in the area. We also updated our vision for the future based on the recent developments in the area.

This book was written to be accessible to practitioners and upper-level undergraduate students. Yet, it has enough in-deep content to be useful for Ph.D. students looking for a topic of research. We expect that this book will be useful for students, researchers, and practitioners interested in Transfer Learning, Reinforcement Learning, Multiagent Systems, and related areas. More than an interesting read, we expect that this book will inspire your research, and help you to see intersections with other communities you would not see otherwise.

Felipe Leno da Silva and Anna Helena Reali Costa
April 2021

Acknowledgments

We wish to acknowledge the collaboration of many colleagues with respect to this book. Be it from conference room discussions, university restaurant chats, or more formal paper reviews, too many people to be explicitly cited had an influence (sometimes anonymously) in this present content. F. L. Silva is especially grateful to Matthew E. Taylor, Peter Stone, and Ruben Glatt for all the career and technical advice, as well as the many hours spent discussing research over the past few years. A. H. R. Costa would like to thank all her graduate students, current and former, for their valuable discussions and contributions over the years.

We would also like to thank Michael Morgan, Christine Kiilerich, Matt Taylor, the anonymous reviewers, and the entire Morgan & Claypool Publishers team involved in the review and editorial process of this book. Without their involvement, this book would probably never go out to the world.

We dedicate this book to our families, who laid the foundation of our education and supported us in many ways over the last years. *Obrigado Paula, Berto e Solange.* A. H. R. Costa would like to thanks her husband Fabio, her mother Rachel, and her children, Regina and Marina. They have helped in so many ways.

Finally, A. H. R. Costa gratefully acknowledges the partial funding from the Brazilian agencies *Brazilian National Council for Scientific and Technological Development* (CNPq) and *São Paulo Research Foundation* (FAPESP). F. L. Silva acknowledges partial funding from *Fundação para o Desenvolvimento da UNESP* (FUNDUNESP - 3061/2019-CCP).

Felipe Leno da Silva and Anna Helena Reali Costa
April 2021

CHAPTER 1

Introduction

Reinforcement Learning (RL) [Sutton and Barto, 2018, Littman, 2015] is a branch of machine learning inspired by conditioning psychology studies on how agents (humans and animals) learn to perform actions in order to maximize rewards that are the observed outcomes of actions.

In RL, an agent learns to reproduce rewarded behaviors and to avoid those penalized. This learning takes place without any feedback on what would be the correct action to maximize rewards. Thus, the agent has to learn by trial and error, through numerous interactions with the environment to discover which actions yield the most reward by trying them.

RL has been employed to train autonomous agents for increasingly difficult tasks such as board and video game playing [Tesauro, 1995, Silver et al., 2016, Vinyals et al., 2019], optimization of treatment policies for chronic illnesses [Shortreed et al., 2011], robotics [Kober et al., 2013], and DRAM memory control [Barto et al., 2017].

As these agents gain space in the real world, it is fundamental, so as to build robust and reliable systems, to learn how to interact and adapt to other (possibly learning) agents. So far, the multiagent RL community has delivered perhaps the most expressive progress toward autonomous learning in Multiagent Systems (MAS) [Bazzan, 2014].

However, the huge sample complexity of traditional RL methods is a well-known hindrance to apply both single- and multiagent RL in complex problems, and the RL community has devoted much effort on additional techniques to overcome the limitations of RL methods.

Transfer Learning (TL) [Taylor and Stone, 2009] methods propose alleviating the burden of learning through the reuse of knowledge. The most intuitive way to apply TL to RL is to reuse the solution of previous tasks that have already been presented to the agent [Taylor et al., 2007]. Many methods have also focused on reusing knowledge from external sources, such as demonstrations from humans or advice from other learning agents [Silva et al., 2017].

The literature has shown that the reuse of knowledge can significantly accelerate the learning process. However, unprincipled reuse of knowledge may cause *negative transfer*, when reusing knowledge hurts the learning process instead of accelerating it. The remaining key question is how to develop flexible and robust methods to autonomously reuse knowledge for varied applications, without incurring negative transfer.

TL for single-agent RL has already evolved enough to consistently reuse solutions from different domains [Taylor et al., 2007], autonomously identify and reuse previous solutions [Isele et al., 2016, Sinapov et al., 2015], and be usable in complex applications.

In contrast, multiagent RL is still struggling to find real-world applications, and many multiagent TL approaches are still validated in toy problem simulations or require strong human intervention. Nevertheless, the field has been maturing in the past years, pushing the boundaries of knowledge closer to the development of an autonomous agent that can learn faster by reusing its knowledge, observing other agents' actuation, and receiving advice from more experienced agents.

This book is an extended survey that aims at categorizing and discussing the main lines of current research within the *Transfer Learning for Multiagent RL* area. Our main objective is to highlight the similarities between those different lines, making it easier to identify crossing points and open problems for which the subcommunities could merge their expertise to work on, bridging the gap between current literature and complex real-world applications.

1.1 CONTRIBUTION AND SCOPE

This book focuses on *Transfer Learning* approaches that are explicitly developed for multiagent RL systems or that can be easily applicable to MAS. While we also discuss some methods that have been primarily developed for single-agent TL, the focus of the discussion is on the benefits and/or difficulties in applying these methods to MAS.

By multiagent RL we mean settings in which more than one autonomous agent is present and at least one of them (the one reusing knowledge) is applying RL. If one agent is solving a single-agent RL task but interference from other agents during learning is possible (such as providing suggestions when the agent is unsure of what to do) [Silva et al., 2020a], we also consider this as a multiagent task. We contribute a new taxonomy to classify the literature in the area into two main categories (as detailed in Chapter 3). We also categorize the main recent proposals and discuss the specificities of each of them, outlining their contribution to the MAS community and prospects of further developments when possible.

Whether or not a transfer procedure should be considered as *multiagent* may prompt debates in some situations. We consider that an agent is composed of its own set of sensors, actuators, and a policy optimizer. Therefore, we included here procedures that: (i) are specialized for reusing the agent's own knowledge across tasks that include multiple agents in the environment; or (ii) transfer knowledge from one agent to another during the learning process. Notice that a human providing demonstrations or guidance during learning is considered a multiagent transfer method, but a designer devising a reward function is not, because the information is defined and made available to the agent before learning.

To the best of our knowledge, this is the first work focused on TL for multiagent RL. Table 1.1 shows the main related surveys in the literature. Taylor and Stone [2009] provide a comprehensive survey on TL techniques (primarily focused on single-agent RL), but many new approaches have arisen since that survey was published. Lazaric [2012] surveys TL approaches at a high level, proposing a taxonomy to divide the field (also focused on single-agent RL). Bignold et al. [2020] survey approaches where RL agents are assisted by external information.

Table 1.1: Summary of main related surveys. Those works either focus on TL or multiagent RL (M-RL), but none focused on the intersection.

Survey	Topic		Focus
	TL	M-RL	
Stone and Veloso [2000]		✓	Multiagent Learning
Busoniu et al. [2008]		✓	Multiagent RL
Taylor and Stone [2009]	✓		General (single-agent) TL
Argall et al. [2009]	✓		Learning from Demonstrations
Lazaric [2012]	✓		High-level TL Categorization
Zhifei and Joo [2012]	✓		Inverse RL
Hernandez-Leal et al. [2019]		✓	Multiagent RL using Deep Learning
Nguyen et al. [2020]		✓	Multiagent RL using Deep Learning
Bignold et al. [2020]	✓		Externally-assisted Agents

Their categorization falls within our definition of TL, but their primary focus is on single-agent tasks. Although Genetic Algorithms [Goldberg, 1989] can generally be used to solve decision-making problems where RL is employed, they are more commonly used in problems where policies can be evaluated quickly and easily, which is often not the case. Therefore, we consider that those methods are out of our scope of this book, unless they are used inside an RL approach.

A handful of surveys focus on multiagent RL without emphasis on TL. Two major examples are the surveys by Busoniu et al. [2008] and Stone and Veloso [2000]. A number of recent surveys focus on the challenges of Multiagent Deep RL [Hernandez-Leal et al., 2019, Nguyen et al., 2020]. While most of the *Learning from Demonstration* and *Inverse Reinforcement Learning* techniques fall within the scope of this survey, Argall et al. [2009] and Zhifei and Joo [2012] already provided comprehensive surveys on those topics. Therefore, here we restrict our analysis to more recent and relevant publications on those areas for avoiding repetitions in discussions already provided by these previous surveys. In this book, we also contribute in-depth discussions of the publications, relating them to other paradigms in *Multiagent Learning*.

This book has extended our prior survey on this same topic [Silva and Costa, 2019]. The preface details in which aspects the survey was expanded into a book format. Figure 1.1 further illustrates our focus, which lies at the intersection of RL, TL, and multiagent learning.

1.2 OVERVIEW

This section explains the structure of this book. In Chapter 2 we present the foundations of single- and multiagent RL, showing how TL improves RL algorithms and presenting some

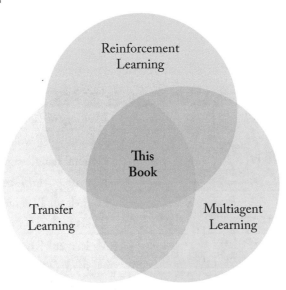

Figure 1.1: Illustration of the scope of this book. Transfer learning for multiagent reinforcement learning lies at the intersection of reinforcement learning, transfer learning, and multiagent learning.

established metrics to evaluate transfer. In Chapter 3, we present our proposed taxonomy, explaining how we group and categorize proposals. In Chapters 4 and 5, we list, explain, and discuss the surveyed published literature, where each of the sections is devoted to one category. Chapter 6 depicts simulated and real-world domains that have been used for evaluating methods. In Chapter 7, we present our view on the promising open questions in the area that need further investigation. Chapter 8 presents pointers for conferences, journals, and code libraries that might be of interest to readers. Finally, Chapter 9 concludes the book.

CHAPTER 2

Background

One of the most primitive ways of learning is by directly interacting with the environment that surrounds us, observing action consequences and evaluating what should be done to achieve our long-term goals.

In order to master a certain behavior, we often repeat it over and over again, seeking to improve it based on trial and error. In this process, we learn the behavior and, in the end, we feel gratified for being able to execute it effectively and efficiently. For example, a baby learns to walk by applying this process: the baby advances and falls. This is repeated many times before the baby masters the art of going quickly from one place to another without getting hurt.

In this type of learning, we do not necessarily have someone instructing us and teaching what should be done continuously. We instead learn through repeated interactions with the environment, being penalized or rewarded according to our successes. This is exactly what characterizes RL. In fact, during this process, we seek to minimize penalties and maximize rewards we receive, learning the best actions to be performed in each situation to achieve this goal.

While RL is a powerful tool to learn how to make the right decisions, certain task characteristics render the learning problem difficult. These characteristics refer mainly to tasks in open, highly dynamic, uncertain, or complex environments, or even with decentralized or distributed data, control, or expertise [Wooldridge, 2009].

Those tasks are typically better solved by multiple agents. In these cases, a solution with multiagent systems can guarantee efficiency and reliability in the execution of tasks, robustness in continuing to perform in face of adversity, flexibility to perform different tasks, and concurrency by carrying out different subtasks at the same time [Bogg et al., 2008].

Multiagent systems are even more powerful when the constituent agents learn from their own experience using RL. In this case, the system is called a multiagent RL system. However, many challenges are imposed to make multiagent RL systems efficient and effective, since the simultaneous actuation of agents makes the environment non-stationary, hampering the learning.

RL also has to cope with other aggravating issues, be it with a single or with multiple agents. A first difficulty is finding how to abstract information and learn different levels of representations, in order to facilitate and scale up the learning process. Deep learning techniques help to overcome this difficulty, as they are typically based on artificial neural networks with the ability to automatically discover the representations needed for better organizing the information received.

Another issue with RL is that the learning process requires many interactions with the environment and, consequently, it takes a long time. Once again inspired by human learning, past experiences in similar situations can be reused to reduce the need for interaction with the environment and to accelerate learning. For example, a child can reuse some of the knowledge and skills learned while riding a tricycle to accelerate her learning to ride a bicycle. In a multiagent scenario, each agent carries its previous experiences and can use this knowledge to accelerate not only its own learning, but also the learning of the entire team. This is the rationale that guides the transfer learning area.

In this chapter, we seek to capture the most important aspects of the problem faced by learning agents interacting over time with their environment to achieve a goal. For this, we review RL fundamentals. We start by addressing the formalization of RL for single-agent scenarios and then expand on that to include learning representations with deep learning. Next, we describe how to solve multiagent scenario. Finally, we present the basic concepts of transfer learning in the context of RL.

2.1 THE BASICS OF REINFORCEMENT LEARNING

RL is concerned with autonomous agents learning successful control policies by experimenting in their environment. The agent has sensors with which it observes the state of the environment. Based on this observation, the agent chooses to perform an action that changes the state of the environment. The task of the agent is defined by a reward function that assigns a numerical value to each action the agent may take in each state. The goal of the agent is to learn a control policy that, from any initial state, chooses actions that maximize the reward accumulated over time by the agent.

RL agents receive indirect, delayed, and occasional rewards. The reward function can be embedded into the learning agent, or known only to an external trainer who provides the reward value for each action performed by the agent. In both cases, the reward indicates the desirability of the resulting situation or state. For example, consider a vacuum cleaner robot that aims to learn how to reach a docking station to recharge its batteries. The task is to learn how to get to the docking station by taking the shorter route from anywhere in the environment. The robot may penalize itself for every action it takes. Thus, in order to minimize penalties, the robot will learn to get to the docking station quickly. The robot must have a sensor that indicates when the docking station has been reached. Another way of modeling this problem would be to give the robot a positive reward whenever it reaches the docking station, and nothing otherwise. This example depicts how indirect, delayed, and occasional the reward can be. The numerical value received by the agent is not directly related to the ideal action to be applied in each state, as would be the case, e.g., for supervised learning. In the first case of our example, whatever the action—worse or better for a given state—the robot will be penalized, and in the latter case, the robot will receive nothing until it reaches the goal. The robot must learn to choose actions that

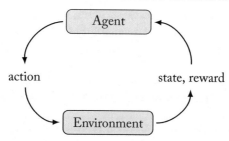

Figure 2.1: An agent interacting with its environment. The agent observes the current state, and chooses an action to perform; after performing the action, the environment evolves and the agent observes the next state and receives a reward. The goal of the agent is to learn an actuation policy that maximizes the expected sum of these rewards.

changes its position, i.e., that alter the state of the environment based on a cumulative reward function defining the quality of any given action sequence.

The agent has also to take into account that observations are sequential and dependent on each other. Moreover, actions affect future observations. Finally, RL occurs through trial and error, via experiences lived by the agent.

To formalize the RL problem, consider an agent trying to solve a task in an environment described by a set of possible states, as illustrated in Figure 2.1. Initially, consider that the problem is formulated with a single agent interacting with the environment. The agent can perform any action from a set of available actions. At each decision step, the observed state provides the agent with all necessary information for choosing an action among the available ones. The agent then decides to perform an action in that state, which causes a state transition. As a consequence, two things happen: the agent receives a real-valued reward, and the environment evolves to a possibly different state at the next decision step. Both the rewards and the transition probabilities depend on the state and the chosen action. The reward indicates the desirability of applying the chosen action in the current state and reaching the next state. As this process evolves through time, the agent receives a sequence of rewards. The goal of the agent is to learn a control policy that, from any initial state, chooses actions that maximize the reward accumulated over time.

More formally, the problem of learning sequential control strategies illustrated in Figure 2.1 is based on a Markov Decision Process (MDP) [Sutton and Barto, 2018, Puterman, 2005]. A stationary discrete-time MDP is a stochastic control process defined as a tuple $\langle S, A, T, R, \gamma \rangle$, where:

- S is a finite set of environment states;

- A is a finite set of available actions to the agent;

- $T : S \times A \times S \to [0, 1]$ is a stochastic state transition function, where the state $s \in S$ is transitioned to state $s' \in S$ with a probability $0 \leq p \leq 1$ when applying action $a \in A$ in state s. Hence $T(s, a, s') = p(s_{k+1} = s' | s_k = s, a_k = a)$, and $\sum_{s' \in S} p(s' | s, a) = 1$;

- $R : S \times A \times S \to \mathbb{R}$ is the reward function. $R(s, a, s') = r_k$ is the immediate reward received after transitioning from state s to state s' due to action a; and

- $\gamma \in [0, 1)$, is the discount factor that represents the relative importance of future and present rewards.

Unless otherwise noted, we use MDP to refer to stationary discrete-time MDP throughout this book.

Small problems usually use atomic representations for states. In an atomic representation, each state is indivisible, with no internal structure; each state is treated as a single element. Large problems are often difficult to solve using atomic representations because identifying commonalities across states becomes hard. In this case, a common and convenient way of describing states is to use a factored representation. In a factored representation, the states are defined by a set of features represented by state variables. Each state is then defined according to the values of the state variables. This structure can be exploited to more efficiently obtain a solution to the problem. There are even more expressive representations, such as the relational representation of states that describes the objects in the world and their relations.

An initial state distribution function defines the probability of starting in each state when the process starts. The set A of allowed actions is defined as $A = \bigcup_{s \in S} A_s$, where A_s is the set of allowed actions in each state $s \in S$, where the model could be restricted to $A = A_s$, $\forall s \in S$. Reward functions describe how the agent should behave and the designer expresses in it what the agent is expected to accomplish. They may be positive or negative, but must be bounded. There are no absolute restrictions on the specification of the reward function, but it can impact the speed of convergence or the success of learning. Reward functions can be defined as $r_k = R(s)$, as $r_k = R(s, a)$, or even as $r_k = R(s, a, s')$ with $R(s, a) = \sum_{s' \in S} r(s, a, s') p(s' | s, a)$. The appropriate definition of the reward function depends on the characteristics of the problem—whether the problem is stochastic or deterministic, whether the state and action spaces are continuous or discrete, whether the rewards are collected continuously or only at the end of the task, etc. Modeling the reward function is one of the most difficult aspects in solving real problems, and impacts directly on how difficult the problem is to solve.

All this process can be triggered when you want to solve a problem or to perform a task. The completion of the task can be defined by terminal states, that is, the task is solved by a sequence of actions that are taken from the initial state until the system reaches an objective state. An episode is defined by the set of all states that are between an initial state and a terminal state. Different episodes are independent of each other. In tasks with no terminal states, the episode is considered to be an infinite episode. RL tasks made of different episodes are called episodic tasks and in these the learning process occurs with the execution of several episodes.

MDPs are solved by observing the current state $s \in S$ and, in this decision step, choosing an action $a \in A$ and executing it. The agent then receives a reward signal r, and observes the next state s' that is obtained as a realization of the probability distribution $T(s, a, s')$. This is repeated until a terminal state or an end condition is reached.

In learning problems, the functions T and R are unknown to the agent, hence a proper actuation must be induced through the observation of $\langle s, a, s', r \rangle$ tuples as gathered through interactions with the environment.

The goal of the learning agent is to learn a way to behave that produces the highest possible cumulative reward over time. A policy π defines how the learning agent behaves at a given time. A stochastic policy $\pi : S \times A \to [0, 1]$ is a mapping from states to probabilities of selecting each possible action. If the agent is following policy π at time k, then $\pi(s, a)$ is the probability that the action at k is a if the state at k is s, i.e., $\pi(s, a)$ defines a conditional probability distribution over $a \in A$ for each $s \in S$. The agent then decides, given $\pi(s, a)$, which action a_k to perform at $s = s_k$. However, in simpler problems, the solution may be a deterministic policy, $\pi : S \to A$, that maps an action to be applied in each possible state, $\pi(s) = a$.

A possible way to learn a good policy is by iteratively updating an estimate of *action qualities* $Q : S \times A \to \mathbb{R}$ after each interaction with the environment. Several algorithms (e.g., Q-learning) are proved to eventually learn the true Q function under non-restrictive assumptions.[1] We define the value Q of taking action a in state s under a policy π as the expected discounted sum of rewards received when starting at s and executing action a, the agent thereafter follows policy π,

$$Q^\pi(s, a) = \mathbb{E}\left[\sum_{j=0}^{\infty} \gamma^j r_{k+j} \,|s_k = s, a_k = a, \pi \right], \tag{2.1}$$

where Q^π is the action-value function for policy π, r_{k+j} is the reward received after j steps from using action a in state s and following policy π on all subsequent steps, and γ is the discount factor that codifies the horizon in which the rewards matter. We consider here the infinite horizon modeling in which the interaction between the agent and the environment does not end after a certain number of steps.

Therefore, an optimal policy $\pi^\star$ is one that maximizes Q in every possible state; in this case, we have an optimal action-value function $Q^\star$ defined as

$$Q^\star(s, a) = \max_\pi Q^\pi(s, a), \tag{2.2}$$

for all $s \in S$ and $a \in A$. Now we can define an optimal deterministic policy,

$$\pi^\star(s) = \operatorname*{argmax}_a Q^\star(s, a). \tag{2.3}$$

---

[1]The full proof for Q-learning is available at Watkins and Dayan [1992]. The main conditions are that: (i) all state-action pairs are infinitely visited; (ii) the rewards are bounded; and (iii) a proper learning rate is chosen.

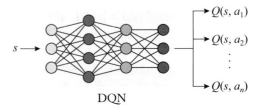

Actions	a_1	$\ldots$	a	$\ldots$	a_n
State s_1					
$\vdots$					
State s			$Q(s, a)$		
$\vdots$					
State s_m					

Figure 2.2: While the Q function is represented by a table in Q-learning (left), in Deep Q-learning it is represented by a deep neural network (right).

Note that we can approximate iteratively the Q-value to $Q^\star$ with each experience $\langle s_k, a_k, s_{k+1}, r_k \rangle$ of the agent. For the Q-learning algorithm, this update rule is

$$Q_{k+1}(s_k, a_k) \leftarrow (1 - \alpha) Q_k(s_k, a_k) + \alpha(r_k + \gamma \max_a Q_k(s_{k+1}, a)), \qquad (2.4)$$

where $r_k = R(s_k, a_k, s_{k+1})$ is the reward, $0 < \alpha \leq 1$ is the learning rate and γ is the discount factor.

Algorithms that follow this procedure are called *value-based RL* because they aim to estimate the value function $Q^\star$ (Equation 2.4 in the Q-learning algorithm) and then, based on this value, define the policy. In value-based RL, no explicit policy is stored, only a value function—the policy is here implicit and can be derived directly from the value function. On the other hand, in *policy-based RL* we explicitly build a representation of a policy and keep it in memory during learning. *Actor-critic* is a mix of the two—these algorithms have in memory an explicit representation of the policy, independent of the representation of the value function. The representation of the policy is called *actor* because it is used to select actions, and the estimate of the value function is called *critic* because it criticizes the actions performed by the actor.

2.2   DEEP REINFORCEMENT LEARNING

So far we have assumed that estimates of value functions are represented as a table. For example, a Q-function has one entry for each state-action pair, which restricts its use only to tasks with a small number of states and actions (see the left side of Figure 2.2).

RL methods must be capable of *generalization* due to the large amount of memory needed to store large tables and the computational time to fill them accurately. Generalization is achieved with the use of any of the existing methods for supervised-learning function approximation, taking each value function update as a training example. This class of algorithms uses a parameterized function approximator with parameter vector θ to represent the value function, instead of using a table to represent it. For example, Q might be the function computed by an artificial neural network, with θ the vector of connection weights.

In fact, artificial neural networks are the function approximators most commonly used in RL. Artificial neural networks with many layers of neurons excel in *deep learning* [Goodfellow et al., 2016], and they are part of a broader family of machine learning methods that perform representation learning. The word "deep" in deep learning refers to the use of multiple layers in the network.

Deep learning progressively extracts higher-level features from the raw input—it can learn by its own which features to optimally place in which level. The feature extraction step is an implicit part of the learning process that takes place in an artificial neural network. During the training process, this step is also optimized by the neural network to obtain the best possible abstract representation of the input data. This characterizes representation learning in deep learning.

Deep neural networks seek to reduce, for each input, the error of the network's output in relation to the target value provided for the training. The error is modeled as a loss function. Since the loss depends on the network's weights, we must find a certain set of weights for which the value of the loss function is as small as possible. The minimization of the loss function is achieved by a *gradient descent method*, in which we gradually move down the negative gradient toward the optimal weights. In particular, backpropagation methods for multilayer neural networks are widely used methods for nonlinear gradient descent function approximation.

RL agents can now be trained using a deep neural network as a function approximator for the value function. In this case, the loss function can be modeled as a quadratic error given by

$$\mathcal{L}(\theta) = \mathbb{E}[\, \| \, (r + \gamma \max_{a'} Q_k(s', a'; \theta)) - Q_k(s, a; \theta) \, \|^2 \,], \tag{2.5}$$

with $(r + \gamma \max_{a'} Q_k(s', a'; \theta))$ being the target. However, combining deep learning with RL poses some challenges that need to be properly addressed.

In supervised learning, the inputs must be independent and identically distributed (i.i.d.) so that the model does not overfit. In addition, the target (i.e., the desired output label in the training set) for a given input should not change over time. This stable condition for each input and output pair enables a good performance for supervised learning.

However, as stated before, in RL not only the observations are sequential and dependent on each other, but also the actions affect future observations made by the agent. This makes the problem non-i.i.d. In addition, as learning evolves along with the agent's interactions with the environment, the agents becomes more knowledgeable of the real state-action values. This changes the target value in Equation (2.5), making the agent try to learn a mapping for an ever-changing input and output. In this way, RL conflicts with some important supervised learning requirements.

Popular ways to alleviate these problems are the use of *experience replay* and a *target networks* [Mnih et al., 2015]. Experience replay consists of storing in a replay buffer a large number of previous experiences. A mini-batch of samples is then randomly sampled from this buffer to train the deep network, forming a set of input data stable enough for training. As the replay buffer is randomly sampled, the data is independent of each other and closer to i.i.d. Experience replay is responsible for great performance improvements in deep learning combined with RL. The purpose of using a target network is to stabilize the Q-value targets temporarily so that it is not a moving target. The idea is to create two deep networks parameterized by θ^- and θ. θ is updated during training and θ^- assumes values copied from θ. Thus, for an agent experience $\langle s, a, s', r \rangle$, the loss function becomes

$$\mathcal{L}(\theta) = \mathbb{E}[\, \| \, (r + \gamma \max_{a'} \hat{Q}_k(s', a'; \theta^-)) - Q_k(s, a; \theta) \, \|^2 \,]. \tag{2.6}$$

From time to time, the two networks are synchronized, $\hat{Q}_k(., \theta^-) \leftarrow Q_k(., \theta)$.

The Deep Q-learning algorithm (DQN) is the "deep" version of Q-learning [Mnih et al., 2015]. Figure 2.2 outlines the value function representation in the Q-learning algorithm and in the DQN algorithm. Note that the output of the DQN is the value of Q for each possible action $a_i \in A$ for state s given in the input. To determine which action a should actually be taken in the observed state s, DQN calculates:

$$\pi(s) = \arg\max_a Q_k(s, a; \theta). \tag{2.7}$$

Note that the input state $s \in S$ can be discrete or continuous. However, actions are discrete in this approach. Most value-based algorithms are off-policy algorithms, that is, the updated policy might be different from the behavior performed for the gathering experiences.

It is also worth noting that DQN uses a fully connected feedforward neural network. However, there is a research line that aims to equip DQN agents with memory. This need comes from the fact that most real-world applications fail to meet the Markov property since their true states are only partially observable. Consequently, this line of research usually replaces the first fully connected layers with a layer of recurrent neural networks. A pioneering example of this line of research is the Deep Recurrent Q-Network (DRQN), proposed by Hausknecht and Stone [2015].

On the other hand, in policy-based algorithms, instead of learning an intermediary Q-function to extract the policy from, many algorithms directly optimize a stochastic policy $\pi(s, a; \theta)$ using the gradient ascent technique, where θ is the set of parameters defining the policy. If actions are discrete, the output of the neural network is the probability of applying each action to the observed state, $p(a_i|s)$, with $\sum_{a_i \in A} p(a_i|s) = 1$, $\forall s \in S$. The probabilities are calculated at the end of each episode, i.e., after the agent develops a trajectory from its initial state to the end, storing the state, action, and reward at each step. The probabilities are increased when they result from high rewards and decreased otherwise. However, these algorithms can also handle continuous actions. In this case, the probability distributions are considered to be normal distributions given by $\mathcal{N}(\mu, \sigma^2)$, μ being the mean and σ^2 being the variance of the normal distribution. These architectures are outlined in Figure 2.3.

The majority of these algorithms are on-policy, which means they only use data from the most recent version of the policy. TRPO [Schulman et al., 2015] and PPO [Schulman et al.,

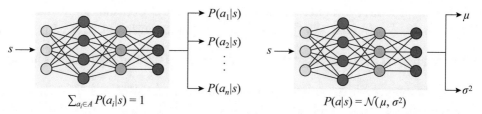

Figure 2.3: Scheme of policy-gradient algorithms. The output of the neural network can be the probability of each discrete action a_i in the observed state s (left) or the probability distribution over actions given by the mean μ and variance σ^2 of a normal distribution.

2017] are some popular examples of such algorithms. The main advantage of DQN over policy-based algorithms is that it is more efficient in terms of sampling than its counterparts. However, DQN does not deal with continuous action spaces.

2.3   MULTIAGENT REINFORCEMENT LEARNING

Although many scenarios and techniques have been discussed so far, MDPs take into account only one agent in the environment, whether they are used in tabular RL approaches or with function approximators. In the multiagent framework, agents act autonomously reasoning over its own knowledge about the world and other agents. In this scenario, each agent has its own utility function that depends on the joint outcome of the actions of all agents. The agents aim at acting strategically in a *multiagent system* (MAS) to maximize utility and achieve their goals.

Of course, it is possible to model all other agents as part of the *environment*, assuming that their policies are fixed. In that way, the local agent learns in a MAS as if it were a single-agent problem [Tan, 1993].

However, generally, the actions of one agent have influence in the local state or reward of the others, making the MDP non-stationary. Thus, agents that act strategically in a MAS should be open to coordination, cooperation, competition, communication, and perhaps negotiation with other strategic agents.

In this context there are two extremes—fully *cooperative* MAS in which agents share the same utility function; and fully *competitive* MAS in which an agent can only win when another one loses. Most agent interactions are between these two extremes and they are often studied in terms of "games." The field of game theory provides a set of analytical tools designed to help us understand the phenomena we observe when decision makers interact [Osborne and Rubinstein, 1994].

The most basic representation of strategic interactions in game theory is the *normal form game*, also known as the strategic form game. A strategic game models a situation in which each agent chooses its plan of action once and for all, and all agents' decisions are made simultaneously [Osborne and Rubinstein, 1994]. A normal form game consists of $\langle N, U, R_{1...n} \rangle$, where:

- $N = \{1, 2, \ldots, n\}$ is the set of agents, and $n = |N|$ is the number of agents in the game;

- U is the joint action space, composed of local actions for all agents, $U = A_1 \times \cdots \times A_n$. An *action profile* u is a tuple $\langle a_1 \ldots a_n \rangle$ that specifies that agent $i \in N$ carries out action a_i, with $a_i \in A_i$; and

- $R_i : U \to \mathbb{R}$ is the reward function of agent i, which returns the expected utility for agent i given the action profile.

Each action profile produces a result on which each agent computes a utility. Utility must represent all the agent's interests, such as justice, altruism, and social welfare [Poole and Mackworth, 2017], just like rewards for RL. The agent seeks to maximize its own utility. In the normal form game the interpretation of an action can be quite general, not only representing a simple choice, but a *strategy* of what the agent will do under the various contingencies (similar to a *policy* in RL).

However, the normal form game does not incorporate any notion of sequence or time of the actions of the agents. Whereas the normal form represents one-shot playing of a game, the extensive (or tree) form is an alternative representation that specifies a particular unfolding of the game in which agents repeatedly play and learn a game. Thus, an extensive form game specifies the possible orders of events; each agent can ponder its plan of action not only at the beginning of the game but also whenever it has to make a decision.

A perfect-information game in extensive form is represented as a (finite) tree in which each internal node represents the choice of one of the agents, each edge represents the action that the agent can perform, and the leaves represent final outcomes over which each agent has a utility function [Osborne and Rubinstein, 1994, Poole and Mackworth, 2017, Shoham and Leyton-Brown, 2009].

In the perfect-information game of the extensive form we assume that the game is fully observable, i.e., at each stage the agents know which node they are at. In an imperfect-information game the state of the world is partially observable. This includes simultaneous action games in which more than one agent needs to decide what to do at the same time. In such cases, the extensive form of a game is extended to include information sets. The agent cannot distinguish the elements of the information set—it only knows that the state of the game is on one of the nodes in the information set, but not on which specific node. Thus, in an extensive form, a strategy specifies a function from information sets to actions.

However, the normal and extensive forms are not always suitable for modeling large or realistic game-theoretic settings. Games may be not finite or they may be played by an uncountably infinite set of agents. The agents may be uncertain about the objective parameters of the environment, or misinformed about the events that happen in the game, uncertain about the actions of other agents that are not deterministic, or even uncertain about the reasoning process of other agents [Osborne and Rubinstein, 1994]. This lead us to Stochastic Games (SG), which are like repeated games but do not require that the same normal-form game is played in each time step. Intuitively speaking, a SG is a collection of normal-form games; the agents repeatedly play games from this collection, and the particular game played at any given iteration depends probabilistically on the previous game played and on the actions taken by all agents in that game [Shoham and Leyton-Brown, 2009]. Notice that we can also see SGs as an extension of Markov decision processes, since the latter are single agent, multiple state models, and the former are multiagent, multiple state frameworks [Busoniu et al., 2008, Bowling and Veloso, 2000].

The reader interested in deepening their knowledge in game theory should not only refer to the textbooks [Osborne and Rubinstein, 1994, Shoham and Leyton-Brown, 2009], but also the vast literature that has evolved over the years.

An SG is composed of $\langle N, S, U, T, R_{1...n}, \gamma \rangle$, where:

- $N = \{1, 2, \ldots, n\}$ is the set of agents, and $n = |N|$ is the number of agents;

- S is the state space. Each state is composed of local states from each agent plus a local state S_0 for the environment (not related to any agent in particular): $S = S_0 \times S_1 \times \cdots \times S_n$;

- U is the joint action space, composed of local actions for all the agents in the multiagent setting: $U = A_1 \times \cdots \times A_n$. Depending on the problem to be solved, the actions of other agents may be visible or not;

- $T : S \times U \times S \to [0, 1]$ is the state transition function, which in multiagent scenarios depends on joint actions instead of local individual actions, and $T(s, u, s') = p(s_{k+1} = s'|s_k = s, u_k = u)$ with $\sum_{s' \in S} p(s'|s, u) = 1$;

- $R_i : S \times U \times S \to \mathbb{R}$ is the reward function of agent i, which is now dependent on the state and joint actions; and

- γ is the discount factor.

The interactions in the SG occur as illustrated in Figure 2.4. At each episode step k, each agent i, $1 \leq i \leq n$, chooses and performs an action depending on its individual policy $\pi^i : S \times A_i \to [0, 1]$. The system evolves from state $s_k \in S$ under the joint action u_k to the next state s_{k+1} that is obtained as a realization of the probability distribution $T(s, u, s')$, and each agent i receives r_k^i as an immediate feedback to the state transition. Akin to the single agent formulation, the goal of each agent is to learn its policy in such a way as to optimize over the received rewards on a long-term perspective.

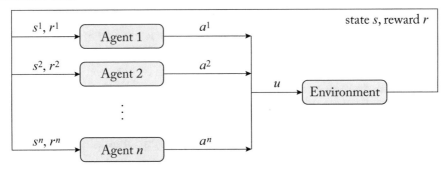

Figure 2.4: In a multiagent setting, each agent observes the current state s_k and chooses an action a_k^i to perform; the environment then evolves as a result of the joint action u_k of the agents and each agent observes the new state s_{k+1} and receives an individual reward r_k^i.

Unlike the case of a single agent, the Q function of each agent i now depends not only on the individual policy of agent i, but also on the policies of other agents, that is,

$$Q^{\pi,i}(s,u) = \mathbb{E}\left[\sum_{j=0}^{\infty} \gamma^j r_{k+j}^i \,|s_k = s, u_k = u, \pi^i \right], \tag{2.8}$$

where $r_k^i = R_i(s_k, u_k, s_{k+1})$ is the reward agent i receives, and γ is the discount factor.

Because each agent i has its own reward function R_i that is dependent on the other agents, there is not a clear definition for optimal policy as in single agent problems. Simply applying the actions that maximize the local reward may be ineffective if the agents have different reward functions, as one agent might (consciously or not) hamper the performance of another. When the policies of other agents are fixed, the agent i can maximize its own utility and find the best policy $\pi^{*,i}$ with respect to the other agents' policies.

We denote the joint policy π as the set of all individual policies, $\pi = \{\pi^1, \pi^2, \dots, \pi^n\}$ and $-i$ denotes all agents except agent i.

Then, $\pi^{-i} = \{\pi^1, \dots, \pi^{i-1}, \pi^{i+1}, \dots, \pi^n\}$. In the situation where one agent cannot improve its own policy when the policies of other agents are fixed, we have a *Nash Equilibrium*, i.e., the best response for each agent i is given by policy $\pi^{*,i}$ so that

$$Q^i_{\pi^{*,i},\pi^{*,-i}}(s,u) \geq Q^i_{\pi^i,\pi^{-i}}(s,u), \ \forall s \in S, u \in U. \tag{2.9}$$

In general, however, when all agents learn simultaneously, the best response may not be unique, and the agents try to learn an equilibrium joint policy [Hu et al., 2015a,b].

Many particular cases can be considered from this more comprehensive formulation.

In a fully cooperative setting, all agents share a single reward $R_1 = \cdots = R_n$ for any state transition [Panait and Luke, 2005]. In such an equally shared reward setting, agents are motivated to collaborate and try to maximize the performance of the team while avoiding the failure of an individual. In this case, it is possible to build a central controller that designates actions for each agent by learning through samples of $\langle s, u, s', r \rangle$. Unfortunately, this solution is unfeasible for most domains due to communication requirements and the huge state-action space for the learning problem. Moreover, with joint reward signs, an individual agent might fail to define the impact of its own action on the team's success, that is, the agent is unable to ascertain its individual contribution to the joint reward received, which makes learning difficult. Associating rewards with agents is known as a credit assignment problem and is a major challenge in this type of scenario.

The *Distributed Q-learning* algorithm [Lauer and Riedmiller, 2000] was proposed to solve such problems in a more scalable way. Each agent learns without observing the actions of the others. However, this algorithm is only applicable in tasks with deterministic transition functions, which is rarely the case for complex tasks.

Equilibrium-based approaches aim at solving the learning problem when agents might have different rewards. For those algorithms, Q-table updates rely on the computation of an equilibrium metric [Hu et al., 2015c]:

$$Q^i_{k+1}(s_k, u_k) \leftarrow (1 - \alpha)Q^i_k(s_k, u_k) + \alpha(r^i_k + \gamma \Phi^i(s_{k+1})), \qquad (2.10)$$

where Q^i_{k+1} is a Q-table related to agent i, α is the learning rate, and Φ^i is the expected equilibrium value in state s_{k+1} for agent i. Equation (2.10) requires the definition of an equilibrium metric, such as the Nash Equilibrium [Hu and Wellman, 2003].

Efficient equilibrium metrics can be extracted using techniques studied in game theory [Sodomka et al., 2013]. The *Learning with Opponent-Learning Awareness* (LOLA) [Foerster et al., 2018a] algorithm follows a similar procedure. Assuming that the other agents in the system are also learning, local policy updates are performed already predicting the policy update of other agents.

Another popular setting is *adversarial learning*, also known as a fully competitive setting, where the agent has an opponent with opposed goals. This problem is described as a zero-sum SG, i.e, the sum of rewards equals zero for any state transition, $\sum_{i=1}^{n} R^i(s, u, s') = 0$. In this case, the optimal policy consists of selecting the action that maximizes the reward supposing that the opponent selected the best action for itself (that is, maximizing the minimum possible return of the actions). For that, the *MinMax Q-learning* algorithm [Littman, 1994] can be used, which updates the Q-table as:

$$Q_{k+1}(s_k, a_k, o_k) \leftarrow (1 - \alpha)Q_k(s_k, a_k, o_k) + \alpha(r_k + \gamma \max_a \min_o Q_k(s_{k+1}, a, o)), \qquad (2.11)$$

where o_k is the action selected by the opponent at step k and o is an action that might be selected by the opponent.

More recently, Lowe et al. [2017] proposed a method especially focused on coordination of multiagent Deep RL problems. In their method, the agents are trained in a centralized setting, where they learn value function estimates taking into account the actuation of other agents. After the training phase, the agents are able to execute the learned policy in a decentralized manner (i.e., using only local observations). Their method was able to handle the non-stationarity of other agents in some problems and achieved convergence where classical solutions failed.

In many MAS applications, each agent might be required to cooperate with a group of agents, while competing against an opposing group. Equilibrium-based solutions are able to generalize to this setting, but it is also possible to treat the opposing team as part of the environment and to learn only how to cooperate with teammates.

Although those solutions solved tasks in some domains, they all require a huge amount of interactions with the environment for achieving a good performance, rendering them hard to scale. Other recent approaches to learn in multiagent RL usually build upon the algorithms discussed in this section to better deal with specific problems, such as non-stationarity [Hernandez-Leal et al., 2017], still maintaining their scalability problems.

Some models and strategies have been specifically proposed to improve in this direction. *Centralized training for decentralized execution* has become increasingly popular in the last years [Foerster et al., 2018b]. During the training process, the agents improve their policies based on information from a centralized critic, but only use decentralized information during execution time. Although this setting is quite restrictive because of the centralized information needed during training time (being only applicable in complex domains if, for example, the training process can be executed in simulation or in controlled environments), the paradigm has become popular because it was a general conception that the centralized view during training would allow learning better execution policies. However, recent research shows that this seems to be a misconception [Lyu et al., 2021] and, in addition to be more restricting than directly learning in a decentralized setting, the final joint policy will have the same performance.

Dec-SIMDPs [Melo and Veloso, 2011] assume that agent interactions only matter in specific parts of the state space, which means that agents must coordinate in a few states contained in S. Agents act as in a single-agent MDP in the remaining states. CQ-learning [De Hauwere et al., 2010] uses the same idea, finding the states in which coordination is needed through a statistical test. Modeling the task with relational representations is also possible to find commonalities and accelerate learning through abstraction of knowledge [Croonenborghs et al., 2005, Silva et al., 2019].

However, one of the most successful strategies for accelerating learning is reusing previous knowledge. In the next section, we formalize and discuss how knowledge reuse is applicable to RL agents.

2.4   TRANSFER LEARNING

Although learning a task from scratch using RL takes a very long time, reusing existing knowledge may drastically accelerate learning and render complex tasks learnable. The area of knowledge transfer has traditionally been called transfer learning (TL). TL algorithms have been very successful and often used to improve the performance of supervised learning algorithms. In this setting, transfer of parameters is a very common procedure. For example, the parameters learned in a neural network that detects cats in images can be transferred to a neural network that detects dogs, reducing the time and difficulty of training the target network.

However, RL has some particularities that allow it to explore TL more broadly. In RL, the number of samples needed to learn a solution from scratch is prohibitive in real-world problems, and leverage the participation of a domain expert to make learning treatable. In addition, every time the task in question changes, the learning process must be restarted from scratch, even when similar problems have already been solved. TL algorithms extract the relevant knowledge from the knowledge collected in solving source tasks and use it to influence the learning process in a similar new task, resulting in a dramatic reduction in the number of samples needed and a significant improvement in learned solution.

Formally, the learning problem consists of mapping a knowledge space $\mathcal{K}$ to a policy $\pi \in \mathcal{H}$, where $\mathcal{H}$ is the space of possible policies that can be learned by the agent algorithm $\mathcal{A}$ [Lazaric, 2012]:

$$\mathcal{A} : \mathcal{K} \rightarrow \mathcal{H}. \tag{2.12}$$

When learning from scratch, $\mathcal{K}$ consists of samples of interactions with the environment. However, additional knowledge sources might be available. One or more agents[2] may be willing to provide further guidance to the learning agent, such as providing demonstrations of how to solve the problem or explicitly communicating models of the problem or environment. Therefore, the knowledge space is often not only composed of samples from the current (target) task $\mathcal{K}^{target}$, but also of knowledge derived from the solution of previous (source) tasks $\mathcal{K}^{source}$ and from communicating with or observing other agents $\mathcal{K}^{agents}$, as illustrated in Figure 2.5. Hence, in the general case

$$\mathcal{K} = \mathcal{K}^{target} \cup \mathcal{K}^{source} \cup \mathcal{K}^{agents}. \tag{2.13}$$

In order to reuse knowledge, it is necessary to decide *when*, *what*, and *how* to store knowledge into $\mathcal{K}$ and reuse it [Pan and Yang, 2010]. Those three questions are hard and long-studied research problems themselves, and there is no single solution valid for all domains. Unprincipled transfer might cause the agent to reuse completely unrelated knowledge, often hampering the learning process instead of accelerating it. This is known as negative transfer, a phenomenon characterized by using knowledge from a source less related to the target task, which can impair the intended performance instead of assisting an effective solution. The literature has investigated many ways to store and reuse (hopefully only) useful information, following varied representations and assumptions.

In general, the main goal of reusing knowledge is to *accelerate* learning. Whether or not a single-agent transfer algorithm learns *faster* than another is commonly evaluated through several of the following performance metrics, summarized by Taylor and Stone [2009] and illustrated in Figure 2.6.

---

[2]Humans or automated agents (actuating or not in the environment) might be involved in knowledge reuse relations.

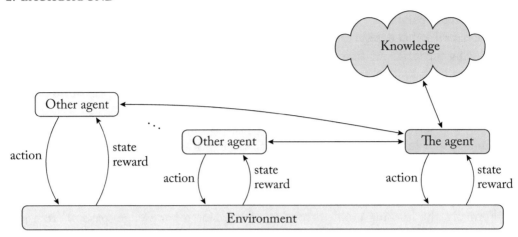

Figure 2.5: Additional sources of knowledge that can accelerate the agent's learning consist not only of samples of the agent's interactions with the environment ($\mathcal{K}^{target}$), but also of information derived from previous experiences that the agent has ($\mathcal{K}^{source}$) or from other agents ($\mathcal{K}^{agents}$), human or not, who may or may not be acting in the environment.

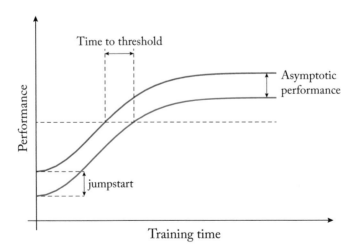

Figure 2.6: Illustration of transfer performance metrics. The blue curve represents non-transfer learning and the red curve represents transfer learning.

The metrics of interest to quantify how much improvement a transfer learning has caused in performance are as follows.

Jumpstart. Measures the improvement in the initial performance of the agent. Since a transfer algorithm might reuse knowledge from $\mathcal{K}^{source}$, successful transfer procedures might initiate learning with superior performance than when learning from scratch.

Time to Threshold. For domains in which the agents are expected to achieve a fixed or minimal performance level, the learning time required to achieve it defines this metric that can be used as a performance metric.

Asymptotic Performance. Agents might be unable to reach the optimal performance in complex tasks, converging to a suboptimal policy. TL might help the agents to reach higher performance, improving their asymptotic performance.

Total Reward. In most transfer settings (either when reusing $\mathcal{K}^{source}$, $\mathcal{K}^{agents}$, or both of them), the transfer algorithm is expected to improve the total reward received during training; this is computed as the area under the reward curve received during the learning process.

Transfer Ratio. The ratio of the total rewards received by two algorithms in comparison; this metric is usually defined in relation to the reward curves received during learning with and without transfer.

While all of these metrics are also generally applicable to MAS, other specific issues must also be taken into account in the multiagent scenario. Communication is a scarce resource in most multiagent applications. Therefore, the *overhead in communication* demanded by the transfer method cannot be ignored, as well as the *computational complexity* and more subjective issues such as the restrictions imposed by the assumptions of the transfer method. For example, if a transfer learning method results in a small improvement in performance by the cost of requiring a much higher communication complexity, could this method be considered as effective?

The trade-off of all those metrics is usually carefully analyzed in a domain-specific manner. The development of better and more comprehensive transfer metrics, for both single- and multiagent, is currently an open subject for research and it is further discussed in Section 7.2.

CHAPTER 3

Taxonomy

TL relies on the reuse of previous knowledge that can come from various sources. Even though an agent could reuse knowledge from multiple sources simultaneously, in practice the current methods usually focus on a single knowledge source. Hence, we here propose dividing the current literature into two main groups in which we can fit all the TL for MAS publications so far. The methods differ mainly in terms of the source of knowledge, availability, and required domain knowledge. We consider the following types of transfer.

Intra-Agent Transfer. This type of transfer refers to the reuse in new tasks or domains the knowledge previously acquired by the agent itself. An intra-agent transfer algorithm has to deal with some challenges such as:

- Which task among the solved ones is appropriate?

- How are the source and target tasks related?

- What to transfer from one task to another?

The optimal answer for those three questions is still undefined. Hence, in practice, usually only one of those aspects is investigated at each time, which means that a real-world application would need to consistently combine methods from several publications. We here characterize intra-agent methods as the ones that do not require explicit communication for accessing internal knowledge of the agents. For example, the procedure of transferring all data from one robot to another similar physical body can be considered as an intra-agent method. However, if this same information is shared with another agent with an identical body, and this agent tries to merge the transferred knowledge with its own experience, then this method qualifies as an *Inter-Agent Transfer*. The publications described in this book are those specialized in multiagent RL, that is, those that assume that there is at least one opponent or teammate in the environment, or those that can be easily adapted for this purpose.

Inter-Agent Transfer. Although the literature on TL for single-agent RL is more closely related to multiagent methods of intra-agent transfer, a growing body of methods focuses on investigating the best way to reuse the knowledge received from communication with other agents, who may have different sensors and possibly different internal representations of what the learning agent in question has. The motivation for this type of transfer is clear: if some knowledge is already available with another agent, why waste time relearning from scratch? However, defining *when*, *what* and *how* to transfer knowledge is not a trivial task, especially if the agents follow different representations. Some methods focus on how to effectively insert human knowledge in automated agents, while others on how to transfer between automated agents. Nevertheless, comprehensive methods would treat any agent equally regardless of their particularities. Some examples of algorithms within this group are the ones derived from *Imitation Learning*, *Learning from Demonstrations*, and *Inverse Reinforcement Learning*. The entity communicating with the local learning agents interplays with the multiagent RL problem in different ways depending on the scenario. If both agents are applying actions and learning the task, then it is just a regular SG scenario where the agents can communicate with each other. If the entity is instead an external observer, who might communicate with the agent but has no influence in the environment, then this scenario can be considered as a kind of partially observable SG. The local agent cannot observe the observer, and the observer's actions affect only the learning agent internal state and not the environment.

In addition to categorizing articles as being intra- or inter-agent transfer, we also classify them in several dimensions, according to their *applicability*, *autonomy*, and *purpose*.

The current literature does not offer a method capable of automatically performing all the necessary steps to transfer knowledge in a MAS (both for intra- and inter-agent transfer). For this reason, most methods focus on a specific subset of problems. Figure 3.1 illustrates how we divide the literature. We use these groupings when discussing the published proposals, and each of them will be explained in more detail in Chapters 4 and 5.

Notice that the classification given here is not rigid, as many methods share properties with more than one of the categories. Moreover, some categories are used for both intra- and inter-agent transfer. In those cases, we group methods based on the focus of their main contribution. Rather than giving a definitive categorization, we here focus on grouping similar approaches to facilitate the analysis of the literature. We list representative papers from recent years in all the surveyed research lines. We also include older papers proposing ideas that are distinct from the recent lines on the literature and, we believe, are not fully exhausted yet.

In the following subsections, we first discuss the nomenclatures used, and then explain all the considered dimensions for paper classification.

3.1   NOMENCLATURE

The literature on knowledge reuse has a myriad of terminologies for closely related concepts. Sometimes, multiple terms have the same meaning or the same term is used inconsistently in different publications. In order to avoid confusion, we here discuss our adopted terminology. While the *jargon* might be inconsistent with the one used in some particular works, we focus on making clear the distinctions and similarities between the discussed methods.

The literature refers to slightly different settings of the knowledge reuse problem under different names.

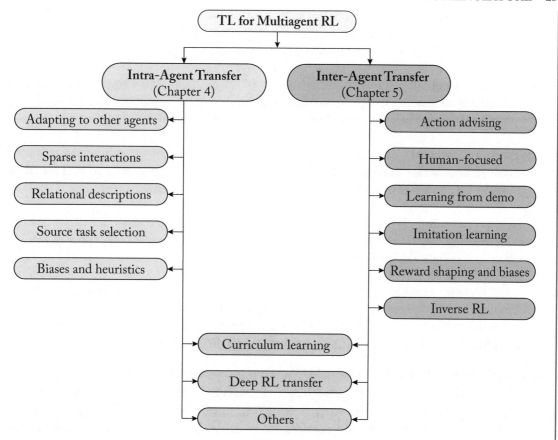

Figure 3.1: Illustration of our taxonomy for the current literature on TL for multiagent RL.

Transfer Learning [Taylor and Stone, 2009] is sometimes used to refer to the problem of reusing knowledge across two or more predefined tasks (the target task has as $\mathcal{K}^{source}$ the knowledge gathered in previous tasks).

Multi-task Learning [Fernández and Veloso, 2006] consists of learning multiple tasks at the same time while exploiting similarities between them to reduce the total learning time ($\mathcal{K}^{source}$ is constantly refined with knowledge from the other tasks).

Lifelong Learning [Thrun and Mitchell, 1995] aims at consistently reusing knowledge across several tasks that might be presented to the agent during its lifespan ($\mathcal{K}^{source}$ grows over time).

Zero-Shot Learning [Isele et al., 2016] tries to reuse knowledge across tasks without training in the target task (i.e., $\mathcal{K}^{target} = \emptyset$).

Learning from Demonstrations [Argall et al., 2009] focuses on the transfer of knowledge between agents where an expert explicitly communicates sequences of actions that solve the task ($\mathcal{K}^{agents}$ contains demonstrations given by other agents).

Imitation Learning [Torabi et al., 2019] occurs when an agent learns to perform a task by observing an expert solving the task without explicit communication. The set of actions used by the agent being observed might even be unavailable to the observing agent.

Inverse Reinforcement Learning [Zhifei and Joo, 2012] consists of an agent trying to learn a task without access to reward samples, and for that, it tries to estimate a reward function, usually by one agent providing samples of a good policy to another agent.

Curriculum Learning [Narvekar et al., 2020] consists of building a sequence of subtasks to the learning agent, in a way that it learns progressively harder tasks easily through reusing knowledge from the previous tasks.

We here use *Transfer Learning* referring to the general challenge of reusing, combining, and adapting knowledge from different sources (agents and tasks).

Regarding the transfer of knowledge between agents, we here adopt the following nomenclatures (first the agent communicating knowledge then the agent receiving it).

Advisor/Advisee: We use this terminology when one agent provides *advice* to another. Here, the knowledge is transferred through explicit communication, but no assumption is made regarding the internal representation or algorithm of other agents. Usually, the *advisor* observes the *advisee*'s state and provides information that is expected to be useful in the current situation (e.g., action suggestions). The advising relation might be initiated either by the *advisee* through a help request or by the *advisor* when the *advice* is expected to be most useful. Optionally, advising might be initiated only when both *advisor* and *advisee* agree on initiating the knowledge exchange.

Teacher/Student: A *teacher* agent also transfers knowledge to a *student* through explicit communication. However, here assumptions about the representation or algorithm might be available. Therefore, more information might be possibly communicated, such as value function estimates, models, rules, or demonstrations of some steps following the optimal policy. The received information might refer to the entire state-action space, and not necessarily only to the current situation.

Mentor/Observer: Here, an *observer* agent tries to imitate successful actuation performed by a *mentor* agent. The *mentor* might be aware or not of the *observer*, but no explicit communication happens. Therefore, the *observer* must use its own sensors to implicitly extract knowledge by observing the *mentor*.

3.2   LEARNING ALGORITHM (LA)

As explained in Chapter 2, many MAS algorithms are specialized to a subset of problems (for example, being applicable only to adversarial settings). As some TL methods are associated with the base RL algorithm, they are susceptible to limitations determined by the RL algorithm. We classify LA in one of the following categories.

Independent ($\mathcal{I}$): Those agents apply single-agent RL algorithms taking into account the local state and actions sets. The other agents and their influence are considered part of the environment dynamics.

Equilibrium ($\mathcal{E}$): Equilibrium-based RL algorithms seek for an equilibrium in the reward functions, usually using game-theoretic approaches to solve the problem. Some TL approaches are based on strategies to reduce the computation of equilibrium metrics, and thus are usable only with these base RL algorithms.

Adversarial ($\mathcal{A}$): Many RL algorithms are specialized to adversarial domains (commonly in a form of zero-sum games). Because of that, some methods have focused on accelerating learning for such adversarial problems.

Cooperative ($\mathcal{C}$): Fully cooperative algorithms assume that all involved agents are benevolent (that is, will never purposely hamper the performance of the system). Therefore, all agents have a common goal. Although not valid for all applications, cooperative algorithms have been broadly used for many problems, especially when the MAS is built by a single owner.

3.3   SOURCE TASK SELECTION (ST)

When the agent already has a library of previous solutions and intends to reuse the gathered knowledge in new tasks, a source task selection procedure must be carried out. Since selecting unrelated source tasks may lead to negative transfer, choosing the correct source task(s) is important to successfully reuse knowledge. The source task selection procedure can be trivially carried out by relying on human intuition to manually select it. However, autonomy is desired for many applications, and human intuition may be deceitful as human "sensors" are different from the agent's. We classify *Intra-agent Transfer* methods in one of the following categories.

Implicit ($\mathcal{X}$): The approach is only usable inside the same domain (e.g., same action and state spaces with possibly varying goals). Therefore, the source task(s) are commonly reused without concerns in regard to negative transfer.

Human-Specified ($\mathcal{H}_s$): The agent considers that the source task is manually selected and given as an input to the transfer method, hence a human user/designer is required to perform source task selection.

Human-Assisted ($\mathcal{H}_a$): A human provides some information to help estimate task similarity (e.g., task parameters). The source task selection is then performed automatically based on this information.

Autonomous ($\mathcal{A}$): The agent estimates task similarity without additional information, and autonomously select the most promising source task(s).

3.4 MAPPING AUTONOMY (MA)

After the source task has been selected, the agent must estimate in which aspects one task is similar to the other. For most applications, applying the entire policy from the source task is either impossible (if the target task has a different state-action space) or will result in a suboptimal actuation. In practice, it is usually more effective to reuse portions of previous solutions but identifying in which aspects the two tasks differ is not a trivial task. We classify the methods in regard to MA in the following.

Implicit ($\mathcal{X}$): The approach is only usable for transfer in the same domain, hence the mapping is straightforward.

Human-Specified ($\mathcal{H}_s$): A human gives an explicit and detailed mapping relating the two tasks.

Human-Assisted ($\mathcal{H}_a$): A human provides some information to help with the mapping. Some examples of such information are relational descriptions or task parameterizations that might help to relate the tasks.

Learned ($\mathcal{L}$): The agent creates an automatically generated mapping. This mapping is usually created by estimating a model for each source task and comparing them to a similar one in the target task.

3.5 TRANSFERRED KNOWLEDGE (TK)

Defining *what* to transfer from one task to another is not trivial. As the best transferable information depends on the setting, there is not a single transfer method which is valid for all situations. For example, if one agent transfers information to another and they have different representations, transferring internal structures may be ineffective or impossible, but if the system is composed of homogeneous agents, this same transfer method may be effective. Due to the myriad of settings in which TL has been applied, several possible types of knowledge transfer have been proposed. The most common ways of encoding knowledge are listed here. However, devising different ways of encoding knowledge for specific scenarios would also be possible. We classify surveyed proposals in the following categories.

Action Advice ($\mathcal{A}_a$): If a group of agents has a common understanding of observations and actions, it is possible for them to communicate action suggestions. Usually, one *advisee* agent communicates its observations to an *advisor*, which can communicate one action to be applied by the *advisee* in the environment.

Value Functions $(\mathcal{V}_f)$: If the agent applies a learning algorithm that uses estimates of value functions to derive policies, it is possible to reuse these estimates across tasks or communicate them to another agent. However, value functions are very specialized for the current task and hard to adapt to similar (yet different) tasks.

Reward Shaping $(\mathcal{R}_s)$: *Reward Shaping* consists of modifying the reward signal received from the environment using additional information to make it more informative to the agent. This information can be originated in a previous task or received from another agent (e.g., another signal received from a human supervisor).

Policy (π): Task solutions might be transferred across tasks if they are similar enough. Another alternative is to transfer portions of the policy (often called as *options*, *partial policies*, or *macroactions*) to benefit from similarities on parts of the tasks.

Abstract Policy (π_a): If a relational description is available, an abstracted version of the task solution might be transferred across tasks or between agents. Abstract policies are more general and easier to adapt to possible differences between tasks.

Rules $(\mathcal{R})$: In addition to being human-readable, rules are easily derived from experience by humans. For that reason, rules have been used to transfer knowledge from humans to automated agents. However, autonomously adapting rules to new tasks is not easy.

Experiences $(\mathcal{E})$: RL agents learn through samples of $\langle s, a, s', r \rangle$ tuples. Those samples can be directly transferred between agents or across tasks. The agent might have to adapt those highly-specialized samples to another task and/or set of sensors though.

Models $(\mathcal{M})$: During learning, the agent can build models to predict the behavior of other agents or characteristics of the task (such as transition and reward functions). Those models can be reused in new tasks or transferred to other agents if they are able to understand it.

Heuristics $(\mathcal{H})$: Random exploration (e.g., ϵ-greedy) is very common for RL agents. However, if a heuristic is available, the agent can make use of it to perform a smarter exploration. Heuristics have been extracted from other agents (most commonly humans) and from previous tasks.

Action Set $(\mathcal{A})$: For problems in which a large number of actions is available, learning the subset of relevant actions for a problem and transferring it to another might be an interesting way to accelerate learning.

Function Approximators $(\mathcal{F}_a)$: Building an explicit table listing all possible quality values for the entire state-action state is often infeasible (and impossible if continuous state variables exist). In this case, a function approximator is usually iteratively refined after each interaction with the environment. They also can be reused later for new tasks or communicated to other agents, but again they are highly specialized for the task at hand.

Bias ($\mathcal{B}$): Instead of reusing the whole policy, the agent might bias the exploration of actions selected by the optimal policy in a previous task. Biases are easier to forget because they quickly lose influence if a bad action is selected.

Curriculum ($\mathcal{C}$): For many complex tasks, it might be more efficient to divide the complex task into several simpler ones. If appropriate task decomposition and order are available, the agent might be able to learn faster by applying TL from the simpler to the target task.

3.6 ALLOWED DIFFERENCES (AD)

Assumptions must be made in regard to in which aspects one task may differ from the other. While the simpler TL algorithms assume that the target task is a version of the source task with a bigger state space, ideally the TL method should allow differences in any element of the MDP or SG description and be able to identify in which aspects the two tasks are similar. In practice, how to identify task similarities is still an open problem, hence we classify the proposals in the following categories.

Same Domain ($\mathcal{S}$): Assumes that the target and source tasks have roughly the same difficulty and are in the same domain (for example, a navigation domain in which only the goal destination can change).

Progressive Difficulty ($\mathcal{P}$): Assumes that the target task is a harder version of the source task in the same domain, usually with a bigger state-action space due to the inclusion of new objects in the environment, but without significant changes in the transition and reward functions.

Similar Reward Function ($\mathcal{R}_f$): Assumes that the reward function remains constant or that the optimal policy in the source task still achieves a reasonable performance in the target task. Unlike in *same domain*, here the target task might have a bigger state-action space, as well as different state variables or actions (possibly requiring mappings).

Any ($\mathcal{A}$): Any aspect of the task might possibly change. The agent must autonomously assess the similarities and discard the source tasks that would result in a negative transfer.

Table 3.1 summarizes all the abbreviations to be used hereafter.

Table 3.1: Quick-Reference legend for abbreviations used in Tables 4.1, 4.2, 5.1, and 5.2.

Learning Algorithm (LA)		Source Task Selection (ST)	
$\mathcal{I}$	Independent	$\mathcal{X}$	Implicit
$\mathcal{E}$	Equilibrium	$\mathcal{H}_s$	Human-Specified
$\mathcal{A}$	Adversarial	$\mathcal{H}_a$	Human-Assisted
$\mathcal{C}$	Cooperative	$\mathcal{A}$	Autonomous
Mapping Autonomy (MA)		**Allowed Differences (AD)**	
$\mathcal{X}$	Implicit	$\mathcal{S}$	Same Domain
$\mathcal{H}_s$	Human-Specified	$\mathcal{P}$	Progressive Difficulty
$\mathcal{H}_a$	Human-Assisted	$\mathcal{R}_f$	Similar Reward Function
$\mathcal{L}$	Learned	$\mathcal{A}$	Any
Transferred Knowledge (TK)			
$\mathcal{A}_a$	Action Advice	$\mathcal{V}_f$	Value Functions
$\mathcal{R}_s$	Reward Shaping	π	Policies
π_a	Abstract Policies	$\mathcal{R}$	Rules
$\mathcal{E}$	Experiences	$\mathcal{M}$	Models
$\mathcal{H}$	Heuristics	$\mathcal{A}$	Action Sets
$\mathcal{B}$	Biases	$\mathcal{F}_a$	Function Approximator
$\mathcal{C}$	*Curricula*		

CHAPTER 4

Intra-Agent Transfer Methods

In this chapter, we look at *Intra-Agent* transfer methods. Tables 4.1 and 4.2 summarize the main publications discussed, which are organized into the following categories: adapting to other agents; sparse interactions; relational descriptions; source task selection; biases and heuristics; curriculum learning, Deep RL transfer; and others. As discussed in Section 1.1, our definition of MAS includes systems in which an agent (possibly human) is communicating with the learner and influencing its learning process, even if that agent is not acting directly in the environment.

For readers interested only in the case where multiple agents are applying actions simultaneously in the environment, we mark with ∗ publications that have only one agent directly changing the environment in their experimental evaluation. Note, however, that many of these proposals are directly applicable in MAS or would need straightforward adaptations.

In the next sections, we group the proposals by their main contributions, and, for all the groups discussed, we provide an overview of the current literature followed by a detailed description of the representative proposals.

4.1 ADAPTING TO OTHER AGENTS

Learning in MAS requires acquiring the ability to coordinate with (or adapt against) other agents. Depending on the task to be solved, agents might have to collaborate with teammates following unknown (and possibly adaptable) strategies, or the environment might allow additions or substitutions of agents at any time [Stone et al., 2010]. Nonetheless, the agent still has to cope with the diversity of strategies assumed by other agents and learn how to coordinate with each of them. Therefore, some TL methods focus on reusing experience for learning how to coordinate with new agents faster. In this section, we discuss representative methods on how previous knowledge can be leveraged for accelerating coordination with other agents.

As shown in Table 4.1, most of the works in this class use adversarial algorithms (**LA**= $\mathcal{A}$), have implicit selection of source task (**ST**= $\mathcal{X}$), perform mapping between tasks assisted or specified by humans (**MA**= $\mathcal{H}$, with $\mathcal{H} = \mathcal{H}_s$ or $\mathcal{H}_a$), transfer policies between tasks (**TK**= π), and require that the source and target tasks share the same application domain (**AD**= $\mathcal{S}$). Very rarely, as in the case of Kelly and Heywood [2015], the agent is able to adapt to new tasks, but they are still assumed to be very similar to previous tasks. The main reason for that is the difficulty of the current literature on identifying when and how the strategy of other agents changed, which makes unfeasible to simultaneously account for significant changes in the task.

Table 4.1: Summary of main recent trends of *Intra–Agent Transfer* methods. We follow the symbols introduced in Chapter 3 and compiled in Table 3.1 for quick reference. The publications are presented in chronological order within their group. Papers denoted with * consider settings where only one agent is directly affecting the environment.

Reference	LA	ST	MA	TK	AD
Adapting to Other Agents (Section 4.1)					
Banerjee and Stone [2007]	$\mathcal{A}$	$\mathcal{X}$	$\mathcal{H}_a$	$\mathcal{V}_f$	$\mathcal{R}_f$
Barrett and Stone [2015]	$\mathcal{C}, \mathcal{A}$	$\mathcal{X}$	$\mathcal{L}$	π	$\mathcal{S}$
Kelly and Heywood [2015]	$\mathcal{I}$	$\mathcal{H}_s$	$\mathcal{H}_s$	π	$\mathcal{R}_f$
Hernandez-Leal and Kaisers [2017]	all	$\mathcal{X}$	$\mathcal{H}_s$	π	$\mathcal{S}$
Hou et al. [2019]	$\mathcal{A}$	$\mathcal{X}$	$\mathcal{X}$	$\mathcal{M}$	$\mathcal{S}$
Sparse Interaction Algorithms (Section 4.2)					
Vrancx et al. [2011]	$\mathcal{E}, \mathcal{A}, \mathcal{C}$	$\mathcal{X}$	$\mathcal{X}$	$\mathcal{R}$	$\mathcal{P}$
Hu et al. [2015a]	$\mathcal{E}$	$\mathcal{X}$	$\mathcal{L}$	$\mathcal{V}_f$	$\mathcal{S}$
Zhou et al. [2016]	$\mathcal{E}$	$\mathcal{X}$	$\mathcal{H}_s$	$\mathcal{V}_f$	$\mathcal{P}$
Relational Descriptions (Section 4.3)					
Proper and Tadepalli [2009]	$\mathcal{C}$	$\mathcal{X}$	$\mathcal{H}_a$	$\mathcal{F}_a$	$\mathcal{P}$
Koga et al. [2013]*	$\mathcal{I}$	$\mathcal{X}$	$\mathcal{H}_a$	π_a	$\mathcal{S}$
Freire and Costa [2015]*	$\mathcal{I}$	$\mathcal{H}_s$	$\mathcal{H}_a$	π_a	$\mathcal{S}$
Koga et al. [2015]*	$\mathcal{I}$	$\mathcal{X}$	$\mathcal{H}_a$	π_a	$\mathcal{S}$
Silva and Costa [2017a]*	$\mathcal{I}$	$\mathcal{H}_s$	$\mathcal{H}_a$	$\mathcal{V}_f$	$\mathcal{R}_f$
Source Task Selection (Section 4.4)					
Sinapov et al. [2015]*	all	$\mathcal{H}_a$	$\mathcal{H}_a$	all	$\mathcal{S}$
Isele et al. [2016]*	$\mathcal{I}$	$\mathcal{H}_a$	$\mathcal{H}_a$	π	$\mathcal{S}$
Braylan and Miikkulainen [2016]	N/A	$\mathcal{A}$	$\mathcal{H}_a$	N/A	$\mathcal{A}$
Biases and Heuristics (Section 4.5)					
Bianchi et al. [2009]	$\mathcal{I}, \mathcal{A}, \mathcal{C}$	$\mathcal{H}_a$	$\mathcal{H}_a$	$\mathcal{H}$	$\mathcal{S}$
Boutsioukis et al. [2011]	$\mathcal{C}$	$\mathcal{H}_s$	$\mathcal{H}_s$	$\mathcal{B}$	$\mathcal{R}_f$
Didi and Nitschke [2016]	all	$\mathcal{H}_s$	$\mathcal{H}_s$	π	$\mathcal{S}$
Curriculum Learning (Section 4.6)					
Madden and Howley [2004]	all	$\mathcal{H}_s$	$\mathcal{H}_a$	$\mathcal{R}$	$\mathcal{P}$
Narvekar et al. [2016]	all	$\mathcal{H}_s$	$\mathcal{H}_s$	$\mathcal{V}_f$	$\mathcal{P}$
Svetlik et al. [2017]*	all	$\mathcal{H}_s$	$\mathcal{H}_s$	$\mathcal{R}_s$	$\mathcal{P}$
Narvekar et al. [2017]*	all	$\mathcal{H}_s$	$\mathcal{H}_s$	$\mathcal{V}_f$	$\mathcal{P}$
Florensa et al. [2017]*	$\mathcal{I}$	$\mathcal{X}$	$\mathcal{H}_s$	$\mathcal{M}$	$\mathcal{P}$
Pinto et al. [2017]*	$\mathcal{I}$	$\mathcal{X}$	$\mathcal{X}$	$\mathcal{C}$	$\mathcal{S}$
Silva and Costa [2018]	all	$\mathcal{H}_a$	$\mathcal{H}_a$	$\mathcal{V}_f$	$\mathcal{P}$
Vinyals et al. [2019]	$\mathcal{A}$	$\mathcal{X}$	$\mathcal{X}$	$\mathcal{C}$	$\mathcal{S}$

Table 4.2: Summary of main recent trends of *Intra-Agent Transfer* methods. Second part of Table 4.1.

Reference	LA	ST	MA	TK	AD
Deep RL Transfer (Section 4.7)					
Agarwal et al. [2020]	$\mathcal{C}$	$\mathcal{X}$	$\mathcal{L}$	$\mathcal{F}_a$	$\mathcal{S}$
Devailly et al. [2020]	$\mathcal{C}$	$\mathcal{X}$	$\mathcal{L}$	$\mathcal{F}_a$	$\mathcal{S}$
Ryu et al. [2020]	$\mathcal{C}$	$\mathcal{X}$	$\mathcal{L}$	$\mathcal{F}_a$	$\mathcal{S}$
Others (Section 4.8)					
Sherstov and Stone [2005]*	$\mathcal{I}, \mathcal{A}, \mathcal{C}$	$\mathcal{H}_s$	$\mathcal{H}_s$	$\mathcal{A}$	$\mathcal{R}_f$
Konidaris and Barto [2006]*	all	$\mathcal{H}_s$	$\mathcal{H}_a$	$\mathcal{R}_s$	$\mathcal{R}_f$
de Cote et al. [2016]*	$\mathcal{I}$	$\mathcal{H}_s$	$\mathcal{H}_s$	$\mathcal{M}$	$\mathcal{S}$
Chalmers et al. [2017]*	$\mathcal{I}$	$\mathcal{H}_s$	$\mathcal{H}_a$	$\mathcal{M}$	$\mathcal{S}$

All types of learning algorithms have been explored by this group of methods, and the most common transfer procedure is the reuse of policies for adapting to new agents in the MAS.

The main objective of methods in this category is to identify what remains constant across tasks or teams of agents to know which knowledge to reuse. Banerjee and Stone [2007] propose a TL method to identify task-independent features to support the transfer of value functions. Their method is applied to *General Game Playing* tasks. It consists of building trees relating game-independent features to possible game outcomes. Those trees can then be used to match portions of the state space in the source and target tasks, enabling the reuse of value functions. Their procedure achieved good results when reusing knowledge between crossing-board games, but it is applicable in quite restrictive situations, as the method is only employable when the opponent follows a non-adaptive and fixed policy for all games. In addition, the human designer is responsible for ensuring that the game's features are applicable and have the same semantic meaning in all games.

Similarly, *PLASTIC-Policy* [Barrett and Stone, 2015] deals with the challenge of adapting the actuation of a learning agent to different configurations of teams [Stone et al., 2010], instead of adapting across tasks. *PLASTIC-Policy* assumes that a set of good policies is available to the agent, which must choose the most appropriate one to cooperate with an (initially) unknown team. For that, a set of beliefs is stored and iteratively updated to predict the expected loss of applying each of the policies according to the profile of the current team. Those beliefs are then used to select the most appropriate policy, and they may change over time if a bad policy is selected. Unfortunately, the possibility of improving the initial policy to adapt to the new team is not considered here.

Indeed, the ability of adapting according to the current agent formation is useful in varied domains. The *DriftER* [Hernandez-Leal et al., 2017] method autonomously detects when an opponent changes its strategy in an adversarial setting. For that, a model of the opponent is learned and used for computing the optimal policy against it. Then, the agent keeps track of the quality of predictions. In case the prediction quality is degraded suddenly, it means that the opponent changed its strategy, that is, the agent must recompute the models. Although capable of identifying when the opponent changes, *DriftER* is not able to reuse past policies for new opponents with similar profiles. For that, *Bayes-Pepper* could be used [Hernandez-Leal and Kaisers, 2017]. Past knowledge is reused by first learning a policy when collaborating or competing with one agent, and then reusing this knowledge when interacting with other agents. Although the paper refers to "opponents," the idea can also be used in cooperative or independent scenarios, as the optimal policy depends on the actuation of other agents in most of MAS. *Bayes-Pepper* assumes that the other agents in the system follow a fixed policy, hence learning how to quickly adapt to learning agents with unknown strategy or learning algorithm can be a fertile ground for future work. Moreover, different ways of modeling other agents could be explored [Albrecht and Stone, 2018].

While those methods primarily focused on identifying when or how the learning target changed, others have focused on how to reuse the knowledge. The *SBB* framework [Kelly and Heywood, 2015] proposes a way to adapt previously learned policies to a new task. Several policies are learned by a *genetic programming* algorithm, and each policy is represented by a network of value estimators that individually predict the quality of each action for the current state. Then, those learned policies are transferred to a new task, forming a hierarchical structure that will learn when to "trigger" one of the past policies. Although originally intended to independent agents, this idea could be merged with *DriftER*, building a hierarchical tree of strategies and identifying when each of the strategies should be used against the current opponent or team. However, *SBB* still relies on a hand-coded mapping of state variables and seems to be ineffective to deal with negative transfer.

Finally, the *eTL-P* framework [Hou et al., 2019] learns several models for opponent agents. The learning process for those models is optimized for diversity and behavioral coverage. The opponent models can later be used to match the opponent strategy, and respond with the best counter-policy.

4.2   SPARSE INTERACTION ALGORITHMS

For some tasks, the actuation of other agents affects the local agent only in a portion of the state space. Thus, it is safe to learn as in a single-agent problem when the actions of other agents do not matter, considerably pruning the joint action space. In this section, we discuss TL techniques specially tailored for this setting.

As seen in Table 4.1, this type of TL has been popular for accelerating *equilibrium* algorithms ($\mathbf{LA} = \mathcal{E}$), which are especially benefited from reducing the size of the space in which the equilibrium metric has to be computed. Most of the literature so far has used these algorithms with the assumption of an *implicit* source task selection ($\textbf{ST} = \mathcal{X}$), but we believe that it would be possible to transfer information about sparse interactions if those algorithms are integrated with source task selectors. The most commonly transferred information is *value functions* ($\textbf{TK} = \mathcal{V}_f$), but some of the methods transfer rules ($\textbf{TK} = \mathcal{R}$) defining when the actions of other agents are expected to be irrelevant for defining the optimal policy.

Different ways of identifying when other agents might affect the local reward have been explored in the literature. Vrancx et al. [2011] propose learning when the actions of other agents in the environment can affect the reward of the local agent in an easy version of the target task, so as to transfer rules defining "dangerous" states in which other agents should be taken into account. Some dangerous states are found through statistical tests, and a rule classifier is trained to generalize the identification of those states. Then, the rules are transferred to the target task. No information regarding the policy or value function is transferred.

Similarly, *NegoSI* [Zhou et al., 2016] computes an equilibrium policy only in the dangerous states, following a single-agent policy in the rest of the state space. Here, the convergence to the equilibrium is accelerated by combining the Q-values from a similar situation in the learned single-agent Q-table with Q-values from a simpler version of the multiagent task, manually selected by the designer.

Other ways of computing the dangerous states have also been devised. The *SVFT* framework [Hu et al., 2015a] learns a single-agent task and introduce new agents that can potentially have influence in the local reward. Then, a model of the transition function in the single-agent task is compared to the multiagent one, and the value function is reused in the states in which the local reward function is independent from other agents' actions.

4.3   RELATIONAL DESCRIPTIONS

The use of relational descriptions in RL is known to accelerate and simplify learning by generalizing commonalities between states [Kersting et al., 2004, Diuk et al., 2008]. For that reason, relations between objects might also help to find similarities or to perform mappings between tasks. In this section, we discuss methods that made use of relational descriptions to transfer knowledge.

Table 4.1 clearly reflects that the main reason for using relational descriptions is for being able to apply *human-assisted* mapping autonomy ($\textbf{MA} = \mathcal{H}_a$). The use of relational descriptions also enables the construction of *abstract policies*, which generalize the knowledge obtained and might help to avoid negative transfer—because of this it can be seen that most of the work of this group transfers abstract policies between tasks ($\textbf{TK} = \pi_a$). Most of the surveyed methods use *independent* learning algorithms ($\textbf{LA} = \mathcal{I}$), but multiagent relational descriptions could be used for compatibility with other learning algorithms. Most also require that the source and target tasks share the same application domain ($\textbf{AD} = \mathcal{S}$).

Although abstract policies facilitate finding correspondences across tasks, using them usually reduces the maximum achievable performance in the task. For this reason, *S2S-RL* [Koga et al., 2013] tries to benefit from the best of both worlds by simultaneously learning a concrete and an abstract policy, using a relational task description. At first, the abstract policy is used to achieve a reasonable actuation faster, and after a good concrete policy is learned the agent switches to the concrete reasoning. Their proposal could be useful to MAS, especially for abstracting interactions between agents, first dealing with a new agent using a general abstract policy and later building a specialized strategy for cooperating with that agent in the concrete level. This initial work used the abstract policies only in a single task, but Freire and Costa [2015] later showed that the learned abstract policies could also be successfully reused through different tasks with promising results for TL.

Up until that point, abstract policies could be built and reused individually, but if multiple previous policies were available only one of them should be chosen for reuse. *AbsProb-PI-multiple* [Koga et al., 2015] was then proposed to combine multiple concrete policies learned in source tasks into a single abstract one. A single combined abstract policy was shown to be better than building a library of policies. Although the method was evaluated in single-agent scenarios, it could be applied to MAS if a multiagent relational task description is used such as Multiagent OO-MDP [Silva et al., 2019] or Multiagent RMDP [Croonenborghs et al., 2005].

While those methods made the potential of relational policies clear, *PITAM* [Silva and Costa, 2017a] autonomously computes a mapping for transferring value functions across tasks based on an object-oriented task description. The required description is intuitive to specify and could be easily adaptable to cope with the multiagent case (each type of agent could be a *class* as in Multiagent OO-MDPs).

More restricting scenarios where relational representations thrive have also been explored. Proper and Tadepalli [2009] specialize for MAS, and propose assigning a group of agents to solve collaboratively a subtask, while ignoring all agents assigned to different subtasks. After learning small tasks composed of small groups of agents, this knowledge can be reused to solve assignments of harder tasks containing more agents. The knowledge transfer is performed by copying the weights of function approximators based on relational descriptions to bootstrap the learning process for new assignments.

4.4   SOURCE TASK SELECTION

Before transferring knowledge between tasks, the first step is to select which of the previous solutions will be reused. The selection of source tasks is a very challenging problem, as the agent does not have information on the dynamics of the environment beforehand for computing task similarities. In this section, we discuss approaches for source task selection.

Table 4.1 shows that source task selection has been largely relying on *human-assisted* mappings so far (**MA** = $\mathcal{H}_a$). The task selection is also mostly assisted by humans (**ST** = $\mathcal{H}_a$), and tasks are expected to have common state-action spaces (**AD** = $\mathcal{S}$) or to share manually defined task features. Source task selectors are commonly independent from knowledge transfer procedures.

Very commonly, mapping procedures are based on designer-specified task features, which indicate what changes from one task to another. Following this trend, Sinapov et al. [2015] propose training a regression algorithm to predict the quality of transfer between tasks within the same domain. The regression is made possible by those task features. When a new task is given to the agent, a transfer quality is predicted according to the previously defined task features, and a ranking of source tasks is built, which can be used for transfer through any *Intra-Agent* method. This approach allowed an appropriate autonomous source task selection without samples of interactions in the target task. However, a human must define appropriate and observable task features. This approach could be directly applied to MAS.

TaDeLL [Isele et al., 2016] also follows this same general idea. When the agent must solve a new task, an initial policy is estimated according to the current value of the task features, and then the initial policy can be iteratively refined. Unlike Sinapov's proposal, this algorithm takes into account the advantages of transferring from consecutive tasks and has constant computational complexity in regard to the number of source policies.

Unfortunately, the paramererization of the current task might not be available to the agent beforehand, hence some methods follow different approaches to map across tasks. Braylan and Miikkulainen [2016] evaluate the policies in the library to select the top performers in the new task and then build an ensemble of source tasks to be used and refined in the target tasks. Although their method was applied only to state transition function estimation, similar ideas could also be used for reusing additional knowledge.

Other ways of finding correspondences have also been explored. Although not directly applying the method to RL in the original publication, Fitzgerald et al. [2016] propose observing demonstrations from a human *teacher*, and those observations are used for refining a mapping function that relates objects in the source and target tasks. Therefore, their method could inspire mapping procedures based on relational descriptions of MDPs.

4.5   BIASES AND HEURISTICS

When learning from scratch, agents must rely on random exploration for gathering knowledge about the environment and task. However, previous knowledge can be used for guiding the exploration of a new task, hopefully increasing the effectiveness of exploration. Biases and heuristics [Bianchi et al., 2015] are two successful ways for guiding exploration.

This class of work varies a lot in relation to the characteristics illustrated in Table 4.1. In this section, we discuss multiagent variants of those approaches.

CB-HARL [Bianchi et al., 2009] relies on building heuristics for supporting the exploration of the environment from a case base built from previously solved tasks. Cases similar to the target task are selected through a similarity function, then the value functions of the previous tasks are transformed into heuristics for exploration directly added to Q-values. *CB-HARL* accelerates learning and is applicable to multiagent tasks, requires similarity functions to compare tasks that demand significant domain knowledge. Therefore, one option for future work could be defining better ways for retrieving tasks from the knowledge base.

Others have leveraged heuristics by only indirectly reusing previous value functions. *BITER* [Boutsioukis et al., 2011] uses a human-specified inter-task mapping to relate the state-action space between tasks. This mapping can then be used to initialize the new policy by biasing it toward the best actions in the source task. This strategy was implemented by introducing a small *bias* value in the Q-values of the action with the highest value function in the source task, prioritizing those actions during the learning process. The advantage of using the bias value is that, in case an incorrect mapping is given, bias values are easy to forget. However, *BITER* requires a very detailed human-coded mapping, which is often unavailable for complex tasks.

Heuristics can also be used to accelerate learning for heavily-parameterized methods. Didi and Nitschke [2016] propose adapting a policy from a previous task to improve the optimization of the NEAT [Stanley and Miikkulainen, 2002] algorithm in the target task. This algorithm codifies the agent policy through a neural network, that has its topology and weights optimized through interactions with the environment. Their method achieved good results in the challenging *Keepaway* domain but is very reliant on a human to define parameters and mappings between the tasks. Therefore, defining a more autonomous way to transfer NEAT-learned policies could be a promising line for further work.

4.6   CURRICULUM LEARNING

Table 4.1 shows that the main characteristic of this group of methods is transferring knowledge across progressively harder tasks ($\mathbf{AD}= \mathcal{P}$). The main focus is usually on how to define task sequences, rather than on how to reuse the knowledge.

Narvekar et al. [2016] borrowed the idea of building a *curriculum* from the supervised learning area [Bengio et al., 2009]. Their main idea is to learn a complex task faster by building a sequence of easier tasks within the same domain and reusing the gathered knowledge across tasks. For that, the designer provides to the agent an environment simulator, which can be freely modified by varying some feature values, and heuristics to guide the agent on how the simulator parameters should be changed to simplify the task. The authors showed that the use of a curriculum can expressively accelerate the learning process. However, in complex tasks, we would have only imprecise simulators to build a curriculum from, especially for MAS in which the policies of the other agents in the system are unknown. Even though it is not clear if the proposal could cope with imprecise simulators, this is nonetheless a very interesting topic for further investigations.

After this seminal work, a number of methods propose ways of building *curricula*, either specialized for specific settings or more general and easier to specify than as in the original paper. Svetlik et al. [2017] propose building a directed acyclic graph, pruning some source tasks that are not expected to help the learning agent according to a *transfer potential* heuristic. Moreover, such a graph allows the identification of curriculum source tasks that can be learned in parallel, which in theory could enable dividing the learning process of the curriculum among several agents (that can combine gathered knowledge using any *Inter-Agent* transfer method). Silva and Costa [2018] later extended Svetlik's proposal by generating the *curriculum* based on an object-oriented description, which could represent a further step toward multiagent *curricula*.

However, despite more flexible than the original *curriculum* paper, those methods still require carefully-specified *curricula* or parameters. With the aim of releasing the designer of the specification burden, Narvekar et al. [2017] proposed autonomously generating a *curriculum*. Their idea is to build a *curriculum* MDP to infer an appropriate task sequence in a customized manner (i.e., taking into account differences in regard to agent sensors or actuators). The *curriculum* MDP agent was able to learn useful *curricula*. However, the amount of learning steps required to learn the *curriculum* is likely to be higher than the number of steps saved when using it.

For this reason, other approaches try to learn a *curriculum* trading-off on flexibility and performance. Florensa et al. [2017] propose changing the initial state distribution to move the task initialization closer to the goal, which causes the agent to receive rewards more quickly, and then progressively start the task farther to the goal to explore the whole environment more efficiently. Although their proposal works in a restrictive setting, this idea could be used for building *curricula* in cooperative MAS, as the agents could, for example, coordinate for starting learning in an initial formation for which the agents do not have much information about.

Others have pursued a similar high-level idea though not following Narvekar's modeling. *Progressive RL* [Madden and Howley, 2004] is an early attempt very similar to *curriculum learning*, where knowledge is progressively transferred from simple to more complex tasks. After learning a task, the agent builds a training set composed of abstracted states (computed through a relational task description defined by the designer) to predict the action suggested by the learned policies. This dataset is then used to train a supervised learning algorithm that learns rules for defining actions for each state. When starting the learning process in a new task, the rules are used for estimating actions in unexplored states. *Progressive RL* considers the possibility of aggregating rules estimated from several tasks and given by humans. However, it cannot cope with inconsistent rules.

RARL [Pinto et al., 2017] models the training process as an adversarial two-player game, indirectly building a *curriculum* to the learning agent. The *protagonist* is the RL learner, which tries to solve the task. On the other hand, the *adversary* causes disturbances in the environment. Both agents apply RL, and the adversary objective is solely minimizing the protagonist's rewards. The authors show that, in this way, the learning agent is able to learn more general and robust policies.

Recently, this type of approach has been part of most of the initiatives solving very challenging domains. *AlphaStar* [Vinyals et al., 2019] uses a *curriculum learning* approach as part of their complex strategy for achieving the Grandmaster level in the challenging StarCraft II game. During the learning process, the agent play games against different versions of itself from the past, and other learning agents specialized in exploiting the learning agent weaknesses. Their approach was shown to be very effective, though very computation and data demanding.

4.7   DEEP REINFORCEMENT LEARNING TRANSFER

Recent developments have built upon function approximations based on deep neural networks. Although many of the ideas discussed through this book are applicable when using such function approximation, some groups have sought ways of building specific network topologies to facilitate knowledge reuse. Due to their ability of learning features from raw data, those function approximators could discover themselves features that facilitate transfer, both across different agents and tasks. In the selected works of this class, we can see in Table 4.2 that they all deal with cooperative algorithms ($\mathbf{LA} = \mathcal{C}$), have implicit selection of the source task ($\mathbf{ST} = \mathcal{X}$), the mapping between tasks is learned ($\mathbf{MA} = \mathcal{L}$), the knowledge transferred are the function approximators themselves ($\mathbf{TK} = \mathcal{F}_a$), and all share the same spaces of states and actions ($\mathbf{AD} = \mathcal{S}$).

Graph Neural Networks (GNN) have been explored by some works to learn embeddings that facilitate transfer. Agarwal et al. [2020] propose a method based on GNNs to train agents in cooperative domains. In their methods, the agents solve partially observable tasks by encoding the environment into entity graphs and parsing them into neural networks that encode the graphs and learn policies in an end-to-end fashion. The network is also equipped with an attention mechanism that learns when to listen to and when to ignore agent communication. All agents share the same network, but act differently in the environment due to the partial observability. The authors showed that the method can successfully reuse knowledge across tasks, and that new agents can be adaptively added and removed from the system.

GNNs have successfully been used by other methods as well. *HAMA* [Ryu et al., 2020] uses GNNs to model the hierarchical relationship between agents (for example agents that cooperate or compete against each other). Those GNNs are then used to learn when to transfer information and how to communicate with other agents. *IG-RL* [Devailly et al., 2020] leverages GNNs to train an agent to control a large network of traffic signs. A single network is trained and the generalization capabilities of GNNs facilitate the inclusion of new agents in the system, as well as to scale up previous models that solved only small instances of this problem.

As Deep RL is likely to continue being the cornerstone of the next novel applications, the development of function approximators specifically optimized for generalization and transfer of knowledge might help to scale up multiagent Deep RL.

4.8   OTHERS

Here we discuss additional methods and ideas that could be potentially useful for multiagent RL but have not been fully explored yet.

While addressing huge state-spaces is a common topic in the literature, some applications might have a huge action-space, hampering the agent exploration. *RTP* [Sherstov and Stone, 2005] identifies a subset of relevant actions in a source task and transfers it to a similar target task. In domains in which the number of available actions is large, properly reducing the number of actions that are considered for exploration presents a huge speed up in the learning process. Although RTP is intended to single-agent problems, it could be easily adapted to work in MAS problems. Reducing the number of actions to be explored could be especially interesting for proposals that learn over *joint* action spaces, which are exponential in function of the number of agents.

As finding correspondences between tasks is one of the major challenges for reusing knowledge, several approaches focus on using or learning representations that enable detecting correspondences easily. Konidaris and Barto [2006] propose learning a value function in the *problem-space* (the regular MDP) while also learning an estimate of the value function in the *agent-space*.[1] The learned function in the *agent-space* is then reused to define a reward shaping, showed to accelerate learning in their paper. Although specialized for single-agent problems, their proposal could inspire the transfer of learned reward shaping functions between agents, for which the main challenge would be to adapt the function to another agent with a possibly different *agent-space*. Chalmers et al. [2017] also take advantage of the separation of the *agent-space* for transferring models that predict consequences of actions. Based on sensors that do not change across tasks, a prediction based on agent-centric information is generated and used for bootstrapping learning if it is estimated to be reliable. The authors report significant speed up in simple tasks, and their idea could be reused for transferring models across tasks for predicting the consequences of joint actions.

While many of the approaches discussed so far are tailored for discrete domains, de Cote et al. [2016] propose a TL method especially tailored for continuous domains. The state transition function is approximated in the source task through a *Gaussian process*. This model is then transferred to the target task, in which some preliminary episodes are executed and a similar model is estimated. Those two models are compared and a new model is generated for regressing the difference between the two tasks. Finally, samples from the source tasks are transformed by using this model, and they are used for bootstrapping learning in the new task. Even though not specialized for MAS in the original proposal, their idea could be reused to quickly adapt to new teammates or opponents by transforming previously observed instances.

---

[1]The agent-space is a portion of the state variables that is always present in any task.

CHAPTER 5

Inter-Agent Transfer Methods

In this chapter, we discuss the main lines of research for *Inter-Agent* methods, which represent the majority of TL techniques for MAS. Tan [1993] presented a comprehensive investigation of simple transfer methods applied to cooperative tasks before the *Transfer Learning* term became widely used. His investigation concluded that agents sharing (i) sensations, (ii) successful episodes, and (iii) learned policies learn faster than agents individually trying to solve a task. However, all those methods have a very high cost of communication to achieve a speed up. As this overhead in communication can quickly render a method unfeasible for most domains, the community has dropped the assumption that communication for sharing knowledge is a sunk cost. Virtually all of recent transfer procedures try to learn the task with limited communication. This has been modeled in different ways, to be discussed in the course of this chapter, including "budgets," communication-aware loss functions, or *teachers* with limited availability. The trade-off learning benefits versus overhead in communication is pervasive in the area.

Tables 5.1 and 5.2 depict proposals that represent current research lines and are organized into the categories presented in Figure 3.1. Notice that those proposals assume that all tasks are in the *same domain* ($\mathbf{AD} = \mathcal{S}$). As the focus in those methods is to extract knowledge from communication with other agents, assuming that this knowledge is specific for a single domain is reasonable for most purposes. For this same reason, most of the literature of *Inter-Agent Transfer* does not perform source task selection ($\mathbf{ST} = \mathcal{X}$) and assume that a trivial mapping of the communicated knowledge is enough ($\mathbf{MA} = \mathcal{X}$). We discuss each group in the next sections.

5.1 ACTION ADVISING

Action advising is one of the most flexible TL methods. Assuming that the agents have a common understanding of the action set and a protocol for communicating information, one agent might provide action suggestions for another even when the internal implementation of other agents is unknown.

In this section, we discuss the main current lines of research on action advising, and, as expected, in this group the knowledge transferred between agents is an action ($\mathbf{TK} = \mathcal{A}_a$). In addition, they all have implicit source task selection ($\mathbf{ST} = \mathcal{X}$) and mapping is also implicit ($\mathbf{MA} = \mathcal{X}$).

Advise [Griffith et al., 2013] is an early advising method. It proposes receiving feedback from a human *advisor* indicating if the *advisee* selected a good action or not. This feedback is not assumed to be perfect, and thus the communicated information is combined with exploration for

Table 5.1: Summary of main recent trends of *Inter-Agent Transfer* methods. We follow the symbols introduced in Chapter 3 and compiled in Table 3.1 for quick reference. Publications are presented in chronological order within their group. Papers denoted with * consider settings where only one agent is directly affecting the environment. For all papers **AD**$= \mathcal{S}$.

Reference	LA	ST	MA	TK
Action Advising (Section 5.1)				
Griffith et al. [2013]*	$\mathcal{I}, \mathcal{A}, \mathcal{C}$	$\mathcal{X}$	$\mathcal{X}$	$\mathcal{A}_a$
Torrey and Taylor [2013]*	$\mathcal{C}$	$\mathcal{X}$	$\mathcal{X}$	$\mathcal{A}_a$
Zhan et al. [2016]*	$\mathcal{C}$	$\mathcal{X}$	$\mathcal{X}$	$\mathcal{A}_a$
Amir et al. [2016]	$\mathcal{C}$	$\mathcal{X}$	$\mathcal{X}$	$\mathcal{A}_a$
Silva et al. [2017]	$\mathcal{I}, \mathcal{E}, \mathcal{C}$	$\mathcal{X}$	$\mathcal{X}$	$\mathcal{A}_a$
Fachantidis et al. [2018]*	$\mathcal{C}$	$\mathcal{X}$	$\mathcal{X}$	$\mathcal{A}_a$
Omidshafiei et al. [2018]	$\mathcal{C}$	$\mathcal{X}$	$\mathcal{X}$	$\mathcal{A}_a$
Ilhan et al. [2019]	$\mathcal{I}, \mathcal{E}, \mathcal{C}$	$\mathcal{X}$	$\mathcal{X}$	$\mathcal{A}_a$
Silva et al. [2020b]*	$\mathcal{I}$	$\mathcal{X}$	$\mathcal{X}$	$\mathcal{A}_a$
Kim et al. [2020]	$\mathcal{C}$	$\mathcal{X}$	$\mathcal{X}$	$\mathcal{A}_a$
Zhu et al. [2020]	$\mathcal{I}, \mathcal{E}, \mathcal{C}$	$\mathcal{X}$	$\mathcal{X}$	$\mathcal{A}_a$
Human-focused Transfer (Section 5.2)				
Maclin et al. [1996]	$\mathcal{I}, \mathcal{A}, \mathcal{C}$	$\mathcal{X}$	$\mathcal{X}$	$\mathcal{R}$
Knox and Stone [2009]*	all	$\mathcal{X}$	$\mathcal{X}$	$\mathcal{R}_s$
Judah et al. [2010]	$\mathcal{I}, \mathcal{A}, \mathcal{C}$	$\mathcal{X}$	$\mathcal{X}$	$\mathcal{A}_a$
Peng et al. [2016a]*	$\mathcal{I}$	$\mathcal{X}$	$\mathcal{X}$	$\mathcal{R}_s$
Abel et al. [2016]*	all	$\mathcal{X}$	$\mathcal{X}$	$\mathcal{A}_a$
MacGlashan et al. [2017]*	$\mathcal{I}$	$\mathcal{X}$	$\mathcal{X}$	$\mathcal{R}_s$
Krening et al. [2017]*	all	$\mathcal{X}$	$\mathcal{X}$	$\mathcal{R}$
Rosenfeld et al. [2017]	all	$\mathcal{X}$	$\mathcal{H}_s$	$\mathcal{F}_a$
Mandel et al. [2017]*	all	$\mathcal{X}$	$\mathcal{X}$	$\mathcal{A}_a$
Learning from Demonstrations (Section 5.3)				
Schaal [1997]*	$\mathcal{I}$	$\mathcal{X}$	$\mathcal{X}$	$\mathcal{E}$
Kolter et al. [2008]*	all	$\mathcal{X}$	$\mathcal{X}$	$\mathcal{E}$
Chernova and Veloso [2009]*	$\mathcal{I}, \mathcal{A}, \mathcal{C}$	$\mathcal{X}$	$\mathcal{X}$	$\mathcal{A}_a$
Walsh et al. [2011]*	all	$\mathcal{H}_s$	$\mathcal{H}_s$	π
Judah et al. [2014]	$\mathcal{I}, \mathcal{A}, \mathcal{C}$	$\mathcal{X}$	$\mathcal{X}$	$\mathcal{A}_a$
Brys et al. [2015a]*	$\mathcal{I}$	$\mathcal{X}$	$\mathcal{X}$	$\mathcal{E}$
Wang et al. [2016]*	$\mathcal{I}$	$\mathcal{H}_s$	$\mathcal{H}_s$	$\mathcal{E}$
Subramanian et al. [2016]*	$\mathcal{I}, \mathcal{A}, \mathcal{C}$	$\mathcal{X}$	$\mathcal{X}$	$\mathcal{E}$
Wang and Taylor [2017]	all	$\mathcal{X}$	$\mathcal{X}$	$\mathcal{E}$
Tamassia et al. [2017]	$\mathcal{I}, \mathcal{A}, \mathcal{C}$	$\mathcal{X}$	$\mathcal{X}$	$\mathcal{E}$
Banerjee et al. [2019]	$\mathcal{C}$	$\mathcal{X}$	$\mathcal{X}$	$\mathcal{E}$
Yang et al. [2020]	$\mathcal{I}, \mathcal{C}$	$\mathcal{X}$	$\mathcal{X}$	π

Table 5.2: Summary of main recent trends of *Inter-Agent Transfer* methods. Second part of Table 5.1.

Reference	LA	ST	MA	TK
Imitation (Section 5.4)				
Price and Boutilier [2003]	$\mathcal{I}$	$\mathcal{X}$	$\mathcal{X}$	$\mathcal{M}$
Shon et al. [2007]	$\mathcal{I}$	$\mathcal{X}$	$\mathcal{X}$	$\mathcal{M}$
Sakato et al. [2014]	$\mathcal{I}$	$\mathcal{X}$	$\mathcal{X}$	π
Le et al. [2017]	$\mathcal{C}$	$\mathcal{X}$	$\mathcal{X}$	$\mathcal{E}$
Torabi et al. [2018]*	$\mathcal{I},\mathcal{A},\mathcal{C}$	$\mathcal{X}$	$\mathcal{L}$	$\mathcal{E}$
Reward Shaping and Heuristics (Section 5.5)				
Wiewiora et al. [2003]*	all	$\mathcal{X}$	$\mathcal{X}$	$\mathcal{R}_s$
Perico and Bianchi [2013]*	all	$\mathcal{X}$	$\mathcal{X}$	$\mathcal{H}$
Devlin et al. [2014]	$\mathcal{C}$	$\mathcal{X}$	$\mathcal{X}$	$\mathcal{R}_s$
Bianchi et al. [2014]	$\mathcal{A}$	$\mathcal{X}$	$\mathcal{X}$	$\mathcal{H}$
Suay et al. [2016]*	$\mathcal{I}$	$\mathcal{X}$	$\mathcal{X}$	$\mathcal{R}_s$
Gupta et al. [2017a]*	$\mathcal{I}$	$\mathcal{X}$	$\mathcal{L}$	$\mathcal{R}_s$
Behboudian et al. [2020]*	$\mathcal{I}$	$\mathcal{X}$	$\mathcal{X}$	$\mathcal{R}_s$
Inverse Reinforcement Learning (Section 5.6)				
Lopes et al. [2009]*	$\mathcal{I}$	$\mathcal{X}$	$\mathcal{X}$	$\mathcal{A}$
Natarajan et al. [2010]	$\mathcal{C}$	$\mathcal{X}$	$\mathcal{X}$	$\mathcal{E}$
Reddy et al. [2012]	$\mathcal{E}$	$\mathcal{X}$	$\mathcal{X}$	$\mathcal{E}$
Shiarlis et al. [2016]*	$\mathcal{I}$	$\mathcal{X}$	$\mathcal{X}$	$\mathcal{E}$
Lin et al. [2018]	$\mathcal{A}$	$\mathcal{X}$	$\mathcal{X}$	$\mathcal{R}$
Cui and Niekum [2018]*	$\mathcal{I}$	$\mathcal{X}$	$\mathcal{X}$	$\mathcal{E}$
Tangkaratt et al. [2020a]*	$\mathcal{I}$	$\mathcal{X}$	$\mathcal{X}$	$\mathcal{E}$
Curriculum Learning (Section 5.7)				
Peng et al. [2016b]*	all	$\mathcal{X}$	$\mathcal{H}_s$	$\mathcal{C}$
Matiisen et al. [2017]*	$\mathcal{I}$	$\mathcal{A}$	$\mathcal{X}$	$\mathcal{C}$
Sukhbaatar et al. [2018]	$\mathcal{I}$	$\mathcal{A}$	$\mathcal{X}$	$\mathcal{C}$
Transfer in Deep RL (Section 5.8)				
Foerster et al. [2016]	$\mathcal{C}$	$\mathcal{X}$	$\mathcal{X}$	$\mathcal{M}$
Sukhbaatar et al. [2016]	$\mathcal{C}$	$\mathcal{X}$	$\mathcal{X}$	$\mathcal{M}$
Devin et al. [2017]*	$\mathcal{I}$	$\mathcal{H}_s$	$\mathcal{H}_s$	$\mathcal{F}_a$
de la Cruz et al. [2019]*	$\mathcal{I}$	$\mathcal{X}$	$\mathcal{X}$	$\mathcal{E}$
Omidshafiei et al. [2017]	$\mathcal{I},\mathcal{C}$	$\mathcal{X}$	$\mathcal{X}$	$\mathcal{E}$
Souza et al. [2019]*	$\mathcal{I},\mathcal{C}$	$\mathcal{X}$	$\mathcal{X}$	$\mathcal{E}$
Lai et al. [2020]*	$\mathcal{I},\mathcal{C}$	$\mathcal{X}$	$\mathcal{X}$	π
Scaling Learning to Complex Problems (Section 5.9)				
Taylor et al. [2014a]	$\mathcal{C}$	$\mathcal{X}$	$\mathcal{X}$	$\mathcal{V}_f$
Kono et al. [2014]	all	$\mathcal{X}$	$\mathcal{H}_a$	$\mathcal{V}_f$
Xiong et al. [2018]	$\mathcal{I}$	$\mathcal{X}$	$\mathcal{X}$	$\mathcal{R}$
Taylor et al. [2019]	$\mathcal{C}$	$\mathcal{X}$	$\mathcal{X}$	$\mathcal{V}_f$

accelerating learning without preventing learning in case of inconsistent feedback. The method achieved good performance when evaluated with a "simulated human." Cederborg et al. [2015] further evaluated *Advise* with human laypeople, and concluded that not only human feedback is useful for learning, but the silence of the *advisor* when observing the *advisee* might be interpreted to extract additional training information.

Then, Torrey and Taylor [2013] propose the *Teacher-Student* framework[1] aiming at accelerating the learning process by receiving advice constrained by limited communication. Their framework also takes into account that the agents may use different internal representations [Taylor et al., 2014a], which means that this framework is usable both for transfer from/to humans and automated agents. Assuming that an *advisor* is willing to provide action suggestions to an *advisee*, they show that it is possible to significantly accelerate learning while constraining communication. *Teacher-Student* can be considered a seminal work, guiding good portion of the posterior action advising works. The literature in this subarea branched in two main groups after Torrey and Taylor's work, based on how each method decides when advice should be given.

The first group leverages handcrafted heuristics to estimate when the advice should happen. Zhan et al. [2016] enabled receiving action suggestions from multiple *advisors*, instead of a single one. Combining multiple advice allowed the *advisee* to avoid bad advice in some situations. Moreover, they also modeled the possibility of refusing to follow advice if the *advisee*'s performance surpasses the *advisor*'s. However, all teachers are still assumed to perform the task at a moderate level of expertise and to follow fixed policies.

Though flexible and allowing the participation of humans, the early frameworks focused on and evaluated advising between automated agents. The *jointly initiated Teacher-Student* framework [Amir et al., 2016] focuses on making this same framework more human-friendly, and accelerates learning while requiring less time observing the *advisee* to provide action suggestions. In this framework, the *advisee* asks for help when unsure of what to do, and then the human *advisor* provides suggestions only if he or she judges that the current state is an important one to provide advice.

However, all those frameworks required an expert agent in the system (human or not) before they could be used. In the *Ad Hoc Advising* framework [Silva et al., 2017], the need of expert *advisors* was finally relaxed. In this framework, agents can advise each other in a MAS composed of simultaneously learning agents. When unsure of what to do, a prospective *advisee* might broadcast an advice requirement, which is answered by *advisors* that have a better policy for the current state. All agents in the environment might assume both roles.

The main weakness of this original formulation is requiring a state counter for computing the agent uncertainty, which hampers applying the method in domains with huge or continuous state spaces. For that reason, both Ilhan et al. [2019] and Silva et al. [2020b] propose ways of computing this uncertainty with Deep RL. The former extends *Ad Hoc Advising* by computing the agent uncertainty with *Random Network Distillation* (RND). The latter computes

---

[1]Note that, despite the original nomenclature, their method fits better in our *advisor-advisee* category.

uncertainties by learning a network architecture that outputs an ensemble of predictions before taking each action. RNDs consist of starting the learning process with two random networks. The first one has fixed weights and the second tries to fit the output of the first network for every state seen by the agent. Therefore, the second network will only match the first one in states that the agent has seen before, and it can be used as a "deep learning version" of a state counter. Silva's approach outputs multiple value predictions for a single state, and just like with RNDs, those predictions will only get close in states that the agent has seen before. Therefore, the variance of those predictions is used as a proxy of the uncertainty.

More recently, Zhu et al. [2020] noticed that the received advice was used a single time, but *advisees* often asked for advice multiple times in the same state, since the knowledge might not be completely assimilated by applying the action a single time. Therefore, they propose reusing previously given advice in their approach. The *advisee* stores the received advice and later reuse it multiple times. The authors show that this procedure might result in a better use of the provided bugdet. However, it is uncertain how their method would perform if only low-performance advisors are available. Therefore, their method should preferably be used when the advisor quality is ensured.

The second group derived from Torrey and Taylor's work has agents learning how to become advisors without human-provided heuristics. Two policies are learned, the task-level policy solves the task, and the teaching-level policy learns the optimal way of providing advice. Zimmer et al. [2014]'s proposal can be considered as a precursor of this line, as they had proposed to build an MDP for learning how to advise.

After Zimmer's initial exploration, *Q-Teaching* [Fachantidis et al., 2018] was proposed based on the notion that the value function of the *advisor* is not always a good guide for selecting the action to be suggested, since the *advisee* will not follow the *advisor*'s policy for a long time. Thus, the horizon (i.e., the discount factor) used to compute the policy might not be valid anymore. By showing that only the performance on solving the task is not enough to choose the best advised action, *Q-Teaching* learns how to give advice, solving an MDP which aims to make the *advisee* learn as fast as possible.

Recently, *LeCTR* [Omidshafiei et al., 2018] was proposed as a more sophisticated method to *learn* when and how to give advice by combining the ideas of those methods and *Ad Hoc Advising*. In *LeCTR*, all agents learn three policies. The first one learns normally how to solve the task, while the other two learn when to *ask for advice* and when to *give advice*. The rewards for improving the advising policies are extracted from metrics estimating the acceleration in the learning process induced by the given advice, and *LeCTL* is able to learn how to provide advice even when the actions have different semantic meanings across agents. Despite the promising initial results, it is still an open question whether *LeCTR* also accelerates learning in complex learning problems.

Similarly, *HMAT* [Kim et al., 2020] decomposes the learning process into learning a "task-level" and a "advising-level" policy. However, here the agents communicate sub-goals that correspond to macroactions, instead of primitive actions. By following this hierarchical approach they show that the team-wide learning process is significantly improved in a toy domain.

5.2   HUMAN-FOCUSED TRANSFER

Although RL algorithms are somewhat based on how autonomous learning is carried out by animals (including humans), actual RL implementations are still very different from the reasoning process of an adult human. Differences on internal representations and how humans and automated agents perceive and interact with the environment impose some obstacles for "transferring" human expertise to an RL agent. Nevertheless, properly receiving guidance and instructions from humans will be required in a world in which machines are integrated into our professional or domestic environments. Moreover, different transfer procedures are usable according to the human technical background. For example, AI experts might be able to design informative reward shapings or rules, while a layperson would probably only be able to give simple instructions, often in a language hard to translate to a machine. Therefore, properly leveraging knowledge from a human is not trivial, and in this subsection, we discuss the main current approaches on this line. Notice in Table 5.1 that diverse information might be transferred from a human to an automated agent, and each method tries to make better use of the costly feedback that a human can give to the agent. Note that they all have implicit source task selection ($\mathbf{ST} = \mathcal{X}$).

RATLE [Maclin et al., 1996] is an early approach to receive human advice in RL. The *teacher* provides advice about actions to take and not to take in certain situations by using a domain-specific language. Then, the instructions are transformed into rules that are later integrated into a neural network that represents the *student* policy. However, their approach requires significant domain knowledge and that the human learns the domain-specific language.

As requiring human teachers to learn a specific language for that might be overly restrictive, later methods have explored how humans can provide feedback in a more intuitive way. The paradigm of observing the *student* actuation and providing feedback about it is the most popular. *TAMER* [Knox and Stone, 2009] is a popular example, where autonomous agents learn through human feedback instead of following the environment reward. The human observes the agent actuation and may provide real-time qualitative feedback such as pressing buttons that mean *good* and *bad*. The agent then learns how to optimize the human shaping reinforcement instead of the reward function. *TAMER* achieved a good-quality policy with a minimal number of training episodes in its experiments. However, as the *student* ignores the environment reward, the learned policy does not improve after the human stops giving feedback.

As a way of enabling exploration to improve upon the demonstrated policy, Judah et al. [2010] divide the learning process into *critique* and *practice* stages. While in the former the *teacher* provides demonstrations by scrolling back the last explored states and labeling actions as *good* or *bad*, in the latter the *student* explores the environment and learns a policy like in regular RL. The final policy is then defined by maximizing the probability of executing good actions while minimizing the probability of executing bad actions. Their approach was validated with humans and presented promising results.

Due to the limited human attention-spam and possible lack of knowledge of how RL agents learn, a number of methods started focusing on ways to make the human teacher indirectly or subconsciously focus on providing feedback in situations where it is expected to be most useful. *SABL* [Peng et al., 2016a] adjusts the velocity of actuation to better guide a human observer over when his or her assistance is most needed (i.e., in what portions of the state-action space there is more uncertainty). When the agent has greater confidence in its policy, the time interval between actions is lower than when confidence is low. The authors have shown that this procedure helps to make better use of limited human feedback, but it is only applicable in environments in which the speed of actuation can be controlled (and in which this does not affect the optimal policy).

Along this same line, *COACH* [MacGlashan et al., 2017] relies on human subconscious feedback to accelerate learning. In their method, the human provides *policy-dependent* human feedback (i.e., the feedback informs how much an action improves the performance of the current policy). As human feedback is often unconsciously based on the observer's performance rather than just on the current selected action, *COACH* showed improvements in a simulated discrete gridworld task and in a real-time robotic task.

Although having knowledge about how the learning process works can help provide better feedback, the human teachers will likely be laypeople in the future, as RL-powered technology becomes more popular. Hence, it is important to facilitate the feedback process. Abel et al. [2016] study how to include a human without knowledge about the *student* reasoning process into the learning loop. This is done by observing its actuation at every step. When the *student* is about to perform a dangerous action, the human *teacher* performs a *halt* operation and introduces a negative reward. The *student* is then expected to avoid executing undesired actions if it is able to generalize this knowledge to future similar situations.

Human teachers might have different levels of technology proficiency. Hence, *SASS* [Rosenfeld et al., 2017] reuses knowledge from humans with technical knowledge on programming and general AI, but without expertise on RL. The human defines a metric that estimates similarities between state-action pairs in the problem space. Although working toward an important challenge of reusing knowledge of non-expert humans, their proposal still needs manual definitions that require expertise in programming, and are probably hard to estimate through simple outputs that could be given by laypeople.

In order to really integrate laypeople in this process, the language for the feedback should be as close as possible from the one they use in their daily life. Krening et al. [2017] propose receiving advice through natural language explanations and translating it to object-oriented advice through sentiment analysis. The result of this procedure is a set of advice that induces the *student* to interact with objects in the environment in a given way and can be used together with object-oriented RL algorithms [Silva et al., 2019, Cobo et al., 2013]. They show that natural language advice can present a better performance than regular learning while facilitating the inclusion of laypeople in the agent training process.

Besides getting feedback complementary to the reward functions, methods can also leverage the human point of view in the task to get structural feedback to improve exploration. *ELI* [Mandel et al., 2017] proposes a different setting for action advising (mainly discussed in Section 5.1). Instead of receiving suggestions of actions to be *applied* in the environment, the *advisee* starts with a minimal set of actions that can be chosen for each state. Then, the *advisee* selects a state for which the available action set is expected to be insufficient, and the *advisor* selects an action to be added to the action set. *ELI* focuses on having humans as *advisors* (in spite of evaluating the method with automated agents) and its main goal is reducing the required human attention during the learning process. In this method, the action chosen by the human is assumed to be optimal, which is not always true in real life.

5.3   LEARNING FROM DEMONSTRATIONS

Learning from Demonstrations is a well-studied category of TL methods in which an experienced *teacher* provides demonstrations to a *student* agent. Demonstrations can be given in many ways, such as teleoperating the *student* or providing some samples of interactions by following the *teacher* policy. The *student* might try to directly learn the *teacher* policy or to use the demonstrations for bootstrapping an RL algorithm. The former achieves good results if the *teacher* policy is close to optimal, but for the general case, the latter is preferred for enabling the agent to learn a better policy than the one used to generate the demonstrations. The main challenge of this category of TL is to leverage the knowledge contained in a limited number of demonstrations, and we here discuss the main current lines of methods. Notice from Table 5.1 that the majority of the discussed methods reuse *experiences* ($\mathbf{TK} = \mathcal{E}$) and that the literature explored all types of learning algorithms. Also, most have implicit source task selection ($\mathbf{ST} = \mathcal{X}$) and implicit mapping ($\mathbf{MA} = \mathcal{X}$).

Titled "Learning from Demonstration," Schaal [1997]'s paper is one of the earliest investigations of this category for both model-free and model-based learning methods. The reuse of demonstrations consisted in simply using the experiences as if the *student* was exploring the environment either by updating value function estimates or refining the model of the task. In his experiments, the model-free experiments achieved only a modest speed up, mostly because the state-action space to explore was much bigger than what was present in a few demonstrations. Model-based algorithms, on the other hand, achieved a relevant speed-up, as the samples from the demonstrations were enough to build a good model of a simple robotic task.

Initially, most of the methods consisted of trying to imitate the *teacher*, without learning by exploring the environment. *HAL* [Kolter et al., 2008] addresses complex RL problems in which the reward signal cannot be observed. For that, the agent receives a hierarchical decomposition of the task from the designer, defining a high-level and a low-level MDP. Then, a *teacher* agent provides demonstrations of an optimal policy, which are used to refine a hierarchical model to estimate the value functions. The parameters for this model are obtained through an optimization formulation that considers the demonstrations at both the high and low level.

Varied algorithms have been used to model the teacher policy. Chernova and Veloso [2007] propose building a model based on *Gaussian mixture models* to predict which action would be chosen by the *advisor* in the current state. In each decision step, the *advisee* estimates its confidence in the prediction for the current state. If the confidence is low, a new demonstration is requested. Even though the authors showed good performance for learning how to solve tasks, their method is not able to improve the *advisor*'s policy through training, which means that the human must be an expert in solving the task. Their later publications improved the confidence estimation procedure [Chernova and Veloso, 2008] by autonomously learning multiple confidence thresholds to ask for advice and added a revision mechanism that allows the advisor to correct wrong decisions [Chernova and Veloso, 2009]. However, the same limitations hold.

Trying to model the teacher policy by analyzing a limited number of demonstrations results in a common problem observed in those early algorithms. The task state-action space is much bigger than what can be explored in demonstrations. Hence, in practice, the student will encounter new situations never seen in the demonstrations. *RAIL* [Judah et al., 2012] (later extended in Judah et al. [2014]) represents an attempt to alleviate this problem. The *advisee* asks for action advice in a sequence of states to learn a classifier that estimates the *advisor*'s policy. The *advisee* alternates between querying the *advisor* and actuating in the environment without asking for help. However, although covering a bigger portion of the task state-action space, the learning agent still cannot improve upon the *advisor*'s policy. Even though the authors explicitly mention that their approach uses fewer demonstrations than Chernova's, they do not focus on reducing the number of interactions with the environment. Therefore, both methods could be better suited depending on the task to be solved.

On its turn, Capobianco [2014] proposes clustering demonstrations given by several different *teachers*, and then using the defined clusters to refine or induce the policy of the *student*. This can be seen as another attempt to have demonstrations covering most of the task, in this case by having multiple teachers.

Therefore, directly imitating the teacher was the prevailing tone of most early learning from demonstration methods. However, as RL progressed to more challenging tasks it became clear that this would be insufficient for many domains. *Human–Agent Transfer (HAT)* [Taylor et al., 2011] is an early example of methods that use the demonstrations as a resource to improve initial performance, but allow the student to explore the environment and improve the policy built from the demonstrations. For *HAT*, a set of rules is learned by analyzing the demonstrations, but the student still learns a policy by using RL. Several ways of deciding how to switch from the rules to the RL policy are evaluated, where considering the rule policy as a possible action seems to be the most efficient in their experimental evaluation.

As learning from exploration is a clearly desired property, most recent methods allow for exploring the environment to improve the demonstrated policy. Walsh et al. [2011] propose learning models for the reward and transition functions from interactions with the environment. Whenever the *student* has a high uncertainty in its predictions, a request for a trajectory of demonstrations is sent to the *teacher*. The authors base their method in the *Knows What It Knows* (KWIK) framework [Li et al., 2011] for presenting theoretical guarantees of minimizing the required number of demonstrations while bounding the "mistakes" in exploration.

Demonstrations have also been used for purposes other than directly building policies based on them. Brys et al. [2015a] propose a method to transform demonstrations into reward shapings. The *student* translates the current state-action tuples in the target task into shaping values by finding the most similar demonstrated states. Then, those values are both used for biasing the initial estimate of value functions and for accelerating value function updates.

Demonstrations have also been reused accross tasks and for supporting a smarter exploration strategy. Wang et al. [2016] propose receiving demonstrations from a human *teacher* in a source task and then reusing those demonstrations in the target task. A simple distance metric is used to translate source samples to the target task, hence both tasks have to share the same state-action space. The proposed method achieved a substantial speed up when combined with a *reward shaping* approach to incentivize returning to frequently visited states. *EfD* [Subramanian et al., 2016] propose using demonstrations for performing a smarter exploration strategy. When exploring the environment, the *student* estimates its confidence in actuating in a given state. In case the confidence is low, the *student* requests demonstrations until a known state is reached. Although efficient, their confidence metric works only when using linear function approximator for value functions. Proposing a more adaptable version of their algorithm could be a promising direction for further work.

As the use of demonstrations became widespread, the main concern of recent works have been both to make an efficient use of the scarce demonstration resource and to concentrate the demonstrations in states where they would be more useful. *CHAT* [Wang and Taylor, 2017] reuses the available demonstrations following a procedure similar to the one applied by Madden and Howley [2004] for transferring knowledge through tasks. First, the demonstrations are used to train a classifier that models the *teacher* policy, which is then reused by the *student* during the learning process. A distinguished feature of *CHAT* is that the classifier algorithm also estimates the uncertainty of the prediction, hence predictions with high uncertainty are quickly dominated by the new policy that is being updated. After some time learning in the task, the classifier is entirely switched to the new policy and not used anymore. This approach was later specialized to MAS by *CC* [Banerjee et al., 2019]. In this approach, a human *teacher* with a global view of the task provides demonstrations containing all local states and actions for all *student* agents. The *students* then use the global information in the demonstrations to estimate the probability of mis-coordinating by following their local observations. This probability is used to define whether to follow a model learned from the demonstrations or to perform the usual exploration in the task.

Demonstrations can also be used to extract information about the task that would be un-available otherwise. Tamassia et al. [2017] propose discovering subgoals from demonstrations given by a *teacher*. Those subgoals are then used for autonomously learned *options*, which accelerate learning. However, their method requires a model of the state transition function of the task, which is often unavailable and hard to learn. Nevertheless, estimating partial policies from demonstrations could be an interesting idea for further investigation.

Recently, demonstrations have been used as a mean to spread global task information in a MAS. *MAOPT* [Yang et al., 2020] assigns to an agent the global view of the environment. This agent learns options based on the local policy of every other agent learning in the environment. After learning suitable options, a set of demonstrations is provided to learning agents, following the learned option. This is equivalent to having one agent providing demonstrations to another, mediated by the centralizer agent that decides when the instruction should begin and end. Although this is an interesting idea for having agents providing demonstrations to each other adaptively, the required bandwidth might make the approach hard to apply in some domains.

5.4   IMITATION

When explicit communication is not available (or other agents are not willing to share knowledge), it is still possible to observe other agents (*mentors*) and imitate their actuation in the environment for acquiring new behaviors quickly. Imitating a *mentor* involves several challenging problems. It might be hard to estimate the performance of another agent online, or maybe even to define which action was applied by the *mentor* (possibly, agents might have different action sets). Finally, imitating another agent might be fruitless if the *mentor* and *observer* have different sensors or learning algorithms in a way that the *observer* is not able to represent that policy. Partly because of those challenges, the literature presents few imitation methods. Table 5.2 shows that those methods extract either *models* or *policies* from observing the mentor. The literature on this group has so far been exploring only domains with independent agents, and most have implicit source task selection ($\mathbf{ST} = \mathcal{X}$).

In order to cope with this challenging scenario, the methods impose restrictions on where they are applicable in. *Model Extraction* [Price and Boutilier, 1999] (later extended in [Price and Boutilier, 2003]) has the *observer* keeping track of the state transitions of a *mentor*. Without any explicit communication and with no knowledge about the applied actions, the *observer* is able to estimate a transition function model, which can be used to update its value function. Although providing a significant speed up in the *Gridworld* domain, the method works in quite restrictive settings. All the agents must have the same state spaces, the action set of the *mentor* must be a subset of the *observer*'s action set, the transition function must be equivalent for all agents, and the local state of each agent cannot be affected by actions of another one (i.e., agents depend only on their local action set, rather than on joint actions).

While one agent trying to imitate another in such restrictive setting would be useful in only a few domains, other works started to explore the role that imitation learning could have in MAS. *Active Imitation Learning* [Shon et al., 2007] proposes a *negotiation* mechanism for imitations, in which agents may "trade" demonstrations of skills useful for them in a game-theoretical way. This method could be especially useful in domains where the agents are self-interested and not necessarily benevolent. Imitation can also be useful in cooperative settings. Le et al. [2017] explores a scenario where multiple *observer* agents try to mimic a team of *mentors*. Assuming that the *mentors* execute a coordinated behavior corresponding to a certain *role* at each time step, a team of *observers* try to learn at the same time an assignment function that defines a role for each agent and a policy for each possible role in the task. Their algorithm iteratively improves a policy for each role given a fixed assignment function, then improves the assignment function given fixed policies for each role. However, their method only mimics observed *mentor* behaviors, and agents are unable to improve the learned policies through exploration.

Several works also solved the imitation problem by reducing it to a Learning from Demonstration scenario (Section 5.3). Sakato et al. [2014] propose an imitation procedure tailored to when an optimal *mentor* is available. The *observer* is divided in two components: the *imitation* and RL modules. The former observes the *mentor* and stores the observed state transition and reward received. Then, the imitation module suggests an action according to the action applied by the *mentor* in a state as similar as possible when compared to the agent's current state. The RL module learns with plain RL. The *observer* selects one of the suggested actions with probability proportional to the similarity between previously observed states and the current one. In their proposal, the agents must have the same action and state sets, and the *mentor*'s actions must be observable. *BCO* [Torabi et al., 2018] represents a contemporary modeling for solving an imitation problem with Learning from Demonstration techniques. Here, the *mentor* demonstrates the solution, but the *observer* cannot initially observe the applied actions (because the agents might have, e.g., different embodiment that would result in different action sets or transition functions). Then, the *observer* acts randomly in the environment for some episodes to learn a model specifying which action would most likely cause the state transition observed at each step performed by the *mentor*. With this model available, the imitation is reduced to a Learning from Demonstration problem, and the authors solve the task using Behavorial Cloning.

5.5   REWARD SHAPING AND HEURISTICS

Exploring the environment is vital for RL agents, as they need to find the best actions in the long-term, which is only possible to compute if all actions have been tried in all states many times. However, the time taken for randomly exploring the environment is usually prohibitive, and a smart exploration procedure is required. For that, it is possible to receive heuristics or reward shapings from *teachers* to improve exploration, hence studying how to build principled heuristics is a very important line of research to make good use of another agent's expertise. Since agents using Potential-Based Reward Shaping (PBRS) are proved to eventually converge to the optimal policy[2] regardless of the shaping quality, those heuristics are some of the few methods that are guaranteed to overcome negative transfer. We discuss the representative methods depicted in Table 5.2 in this section.

The early reward shaping methods laid out the foundation by showing that, not only could the optimal policy learning quickly if good shaping is available, but also that the optimal policy is still learned in case bad shaping is provided. Wiewiora et al. [2003] is one such work, proposing to use look-ahead and look-back reward shaping procedures to incorporate advice in the learning process. Their main insight in this paper is that look-back reward shaping performs better when the advice incorporates preferences over actions, while look-ahead reward shaping has a better performance with state-based preferences.

As the potential of reward shaping became clear, this type of knowledge transfer started to be explored in many different ways. *DIS* [Devlin et al., 2014] is specifically tailored to MAS. The received reward is modified by both a human-given heuristic function and the difference between the system performance with and without each agent, which guides the system to a good joint policy much faster than regular learning. Methods transforming previously learned policies to reward shaping have also been proposed in the single-agent case [Brys et al., 2015b] and could be easily extended to MAS.

Some works have also built shapings based on the policy of more experienced agents. *DRiP* [Suay et al., 2016] uses Inverse Reinforcement Learning methods (see Section 5.6) to derive a reward shaping function used to accelerate learning. First, a set of demonstrations is provided and a reward function is estimated from it. Then, this reward function is used as a potential reward shaping function, which has showed to accelerate learning in challenging tasks such as playing Mario. Gupta et al. [2017a] use a reward shaping function based on how much the agent actuation is corresponding to a behavior demonstrated by an agent with (possibly) different sensors and actuators. For defining those rewards, the first step is to learn an abstraction function that maps the agent concrete state to an abstract space shared with the other agent that will demonstrate the behavior. A neural network is trained to learn a transformation function that can be used to match trajectories generated by the two agents. Finally, the reward shaping value is defined according to how similar the trajectory being followed by the agent is to the demonstrated behavior.

The ability of transforming an arbitrary function into a shaping signal without affecting the original optimal policy from the environment reward function is the main reason why those methods are interesting. Indeed, *PIES* [Behboudian et al., 2020] was shown to accelerate the learning process when a "good" shaping function is available, while guaranteeing that the optimal policy will eventually be learned even when a very bad shaping function is provided.

Methods based on heuristics follow the same high-level idea, though they are usually more open-ended. *HfDARL* [Perico and Bianchi, 2013] uses demonstrations to build a heuristic, later

---

[2]More specifically, to an optimal policy in an MDP [Ng et al., 1999] or the set of *Nash Equilibria* in a stochastic game [Devlin and Kudenko, 2011].

used in the exploration procedure. A spreading procedure is also applied to generalize the usually scarce number of samples, directing the exploration to states spatially close to the ones visited in demonstrations.

Heuristics have been specially useful in adversarial settings. *HAMRL* [Bianchi et al., 2014] is focused on improving exploration in multiagent adversarial tasks. Even simple and intuitive heuristics were shown to be enough to significantly accelerate learning in simulated robotic soccer.

5.6   INVERSE REINFORCEMENT LEARNING

Although most of the RL techniques assume that rewards can be observed after each action is applied in the environment, *Inverse Reinforcement Learning* (IRL) techniques assume that there are no explicit rewards available, and try to *predict* a reward function through other information made available by agents in the environment, such as demonstrations of the optimal policy. However, estimating a reward through observed instances is an ill-posed problem, as multiple reward functions might result in the same optimal policy. Classical approaches for solving this difficult problem suffer from both a high number of required instances and computational complexity [Ramachandran and Amir, 2007], and alleviating those issues is still a topic of current research. As Zhifei and Joo [2012] provide a good summary of those techniques, we avoid the repetition of discussions in their survey and highlight some of the recent research trends of most interest to this survey.

Active IRL [Lopes et al., 2009] alleviates the burden of demonstrating the optimal policy for IRL techniques by making better use of human feedback. For that purpose, the *advisee* predicts its uncertainty in the current reward function to actively ask for an action suggestion in states for which the uncertainty is high. Even though their method requires fewer actions suggestions than simply receiving demonstrations in an arbitrary order, the agent must be able to freely change the state of the task for asking for guidance in the correct states, which is unfeasible for most domains. Still, this method inspired *ARC* [Cui and Niekum, 2018], which moved IRL closer to real applications. The *advisee* generates a trajectory that is expected to maximize the gain of knowledge, according to an uncertainty function. Then, a human *advisor* segments the generated trajectory in "good" and "bad" portions, and both feedbacks can be used to improve the current reward model. Although the computational effort to choose an appropriate trajectory is still prohibitive to solve complex tasks, their method requires less human effort and might inspire researchers in the field.

Despite the fact that most of the classical IRL techniques focus on learning reward functions for single-agent problems, recent proposals have begun to adapt IRL to learn multiagent reward functions. Supposing that the agents follow a *Nash Equilibrium*, Reddy et al. [2012] propose a method for approximating the reward functions for all agents by computing them in a distributed manner.

IRL has been used for purposes other than that as well. Natarajan et al. [2010] deals with problems in which a centralizer agent has to coordinate the actuation of several autonomous agents. IRL is used for estimating the internal reward function of the autonomous agents, and this information is used for improving the policy of the centralizer.

BMIRL [Lin et al., 2018] consists of a Bayesian procedure specialized for two-player zero-sum games. Through a reward prior and an estimated covariance matrix, their algorithm is able to refine an estimated reward function to adversarial tasks. However, their method requires the complete joint policy to be observable, and that the state transition function is completely known, and this information is usually not available for real-world complex tasks.

A common issue of those methods is that the demonstrations are expected to be (at least nearly) optimal. In many domains, the *advisor* might have to repeat an episode multiple times until it is finished successfully. All those failed attempts would then be either wasted or added as undesired noise. Because of that, some recent methods propose ways to use this data. *IRLF* [Shiarlis et al., 2016] integrates failed demonstrations in the learning process. In addition to guiding the agent toward successful demonstrations, their optimization process learns to avoid failed demonstration, reducing the space of possible behaviors that are consistent with the demonstrations. Similarly, *VILD* [Tangkaratt et al., 2020a] can process demonstrations of varying qualities. *VILD* explicitly considers that demonstrators of different expertise levels will not follow the same policy, and provides a way to leverage data from multiple demonstrators.

In general, current IRL techniques are dependent on the full observability of the environment and require high-complexity calculations, which is not realistic for many complex tasks. Yet, IRL is a promising line of research, as training autonomous agents without explicit reward functions might be required for the development of general-purpose robots that receive instructions from laypeople. Despite these two issues, the recent trend on multiagent IRL might inspire new methods for influencing a group of agents to assume a collaborative behavior [Reddy et al., 2012, Natarajan et al., 2010, Lin et al., 2018].

5.7   CURRICULUM LEARNING

As discussed in Section 4.6, a *curriculum* is an ordered list of source tasks intended to accelerate learning in a target task. A promising new research line is the *Transfer of Curriculum*, where one agent builds a *curriculum* and transfers it to another. Preferably, this *curriculum* should be tailored to the agent's abilities and learning algorithm. In this section, we discuss the first investigations on this line. Notice in Table 5.2 that all methods in this groups transfer *curricula* between agents (**TK**$= \mathcal{C}$).

Peng et al. [2016b] study how laypeople build *curricula* for RL agents and show that it is hard for non-experts to manually specify a *curriculum* that will be useful to the learning agent. Since the inclusion of non-experts is fundamental for scaling the applicability of RL agents in the real world, trying to understand and improve how humans instruct automated agents is of utmost importance.

Partially motivated by this negative result on laypeople building *curricula*, other works explored methods where one automated agents builds a *curriculum* to another. *TSCL* [Matiisen et al., 2017] proposes to iteratively build *curricula* by using two agents. The *teacher* solves a *Partially Observable MDP* in which actions represent tasks to present to a *student* and observations are the *student*'s difference in performance since the last time that the same task was chosen. Their experiments report results comparable to a very carefully hand-coded *curriculum* while using much less domain knowledge.

Indeed, the prevailing tone of works in this area is having agents that directly or indirectly tries to learn how to build a *curriculum* by observing what another agents is doing. Sukhbaatar et al. [2018] propose dividing the learning process into two components. A supervisor agent tries to solve the target task for a given time, and then the control is switched to another agent that tries to reverse what the supervisor did (or alternatively reach the same final state) faster than the supervisor. The supervisor can choose when to switch control through an action, which indirectly builds a task to the other agent (the current state and number of steps previously taken by the supervisor affect the task to be presented). When this procedure is repeated, a *curriculum* is built automatically. Although both components are controlled by a single agent in the original publication, implementing each of them in different agents in a MAS could be promising to accelerate the learning process of the system as a whole.

5.8   TRANSFER IN DEEP REINFORCEMENT LEARNING

Most of the recent application successes achieved by RL rely on function approximators to estimate the quality of actions, where *deep neural networks* are perhaps the most popular ones due to their recent successes in many areas of machine learning. Naturally, using Deep RL introduces additional challenges to train the deep networks used for function approximation. Especially, their sample complexity requires the use of additional techniques for scaling up learning (e.g., *experience replay*). Therefore, transfer methods especially focused on Deep RL scenario might help to scale it to complex MAS applications, since the adaptation of this technique to MAS is still in its first steps [Castaneda, 2016, Gupta et al., 2017b]. Two similar investigations concurrently carried out by different groups evaluated the potential of reusing networks in Deep RL tasks [Glatt et al., 2016, Du et al., 2016]. Their results are consistent and show that knowledge reuse can greatly benefit the learning process, but recovering from negative transfer when using Deep RL might be even harder. In this section we discuss other recent *Inter-Agent* methods especially focused on Deep RL.

While harder to train, Deep RL has enabled learning capabilities not available before. It was through this approach allied with transfer that agents could learn how and when to communicate. *DIAL* [Foerster et al., 2016] and *CommNET* [Sukhbaatar et al., 2016] were concurrently proposed as the pioneer approaches to learn how to act simultaneously learning how to communicate. While the former considers discrete communication steps where the gradients can be transferred through agents to accelerate learning, the latter proposes continuous communication cycles and deals with systems with a dynamic number of agents.

The network modularity can also be exploited for devising transfer methods. Devin et al. [2017] decompose the learning problem into two components: a *task*-specific and a *robot*-specific one. They assume that the observations (for practical purposes, the state) can be divided into two sets of state features, which are related to one of the aforementioned components. Furthermore, the cost function can also be decomposed into those two components. For this reason, a modular neural network architecture can be built, where the *task* and *robot* modules might be reused across similar tasks and/or robots. The authors were able to successfully reuse modules in simulated robotic manipulation tasks. However, tasks and robots were very similar and the network architecture was carefully built for their experiments. Additional investigations would reveal if the method works for reusing knowledge across less similar tasks and robots, as well as how to reduce domain-specific parameterizations.

As networks are relatively easy to reuse after switching training modalities, it is also possible to train a network in a task and reuse portions of it in another. Indeed, de la Cruz et al. [2019] propose using demonstrations to pre-train a deep network. At first, the *student* will try to imitate the demonstrations, and later the policy can be refined with regular RL.

Because experience replay mechanisms are present in almost any Deep RL approach, the replay buffer has been used as means for transfer by some methods. *Dec-HDRQN* [Omidshafiei et al., 2017] is an experience replay mechanism focused on multi-task learning in multiagent partially observable tasks. In order to reduce the mis-coordination induced by multiple agents performing experience replay independently, this method defines a unified sampling for building mini-batches by selecting samples corresponding to the same time steps for all agents in the system. Although presenting interesting results, the approach is only applicable if all agents are using similar experience replay and learning mechanisms. *ES* [Souza et al., 2019] leverages the replay buffer typically stored in Deep RL methods to propose a collaborative experience replay mechanism. Whenever multiple instances of the environment can be executed in parallel (e.g., multiple similar robots trying to accomplish the same task), the agents can keep a database of high-quality experiences, and share those among them to learn faster.

Methods originally developed to supervised deep learning can also be used to support TL. *DPD* [Lai et al., 2020] leverages policy distillation to perform TL between two simultaneously learning agents. During the training process, the agents alternate between updating their policies directly from the experiences in the environment, and updating by minimizing the distance between the two policies when the partner's expected return is higher. Their approach achieve significant performance improvements in a variety of control tasks, and might be a good option when multiple instances of the training environment can be executed in parallel.

5.9 SCALING LEARNING TO COMPLEX PROBLEMS

In this section, we discuss papers that focused on solving the learning task in high-complexity domains or had promising ideas that can be implemented in the future. Although not presenting novel transfer approaches, the engineering effort to apply techniques already discussed in this paper for challenging domains might help in the development of new TL techniques.

Complex real-world problems often require balancing between multiple objectives, and having agent with varied level of expertise in the system. *PTL* [Taylor et al., 2014b] delivers a good performance in a real-life problem where this issue is relevant. Using a value function transfer procedure coupled with the *Distributed W-Learning* [Dusparic and Cahill, 2009] multiagent RL algorithm, they accelerate the learning process of a smart grid problem in which multiple electric vehicles aim at recharging their battery without causing disturbances in a transformer. Their transfer procedure relies on sharing value function estimates between agents in a neighborhood, accelerating learning speed at the cost of additional communication. They later investigate varied strategies to decide when to transfer strategies across agents, and how large should be each communication [Taylor et al., 2019]. They also consider different ways of deciding when and how to merge the received message with local knowledge. However, their comprehensive evaluation is carried out in simpler domains.

Most approaches discussed throughout this book consider only TL between pre-defined or slowly changing groups of agents. Few works explicitly consider that, in the real world, many agents (possibly built by different manufacturers) might dynamically participate in TL relations. *KCF* [Kono et al., 2014] can be useful in this situation. This algorithm builds an ontology to transfer knowledge between heterogeneous agents. In Kono et al.'s conception, robots produced by any manufacturer could be linked to a common cloud server giving access to a human-defined ontology tailored for a task. Then, the robot's sensors and actuators would be mapped to an abstract state-action space that would be common to any robot trying to solve that task, regardless of particularities of its physical body. Such mapping would allow different agents to communicate and transfer knowledge. Effectively, any agent could be a *teacher* or *student*, and knowledge would be transferred through the internet. However, in their proposal, an ontology comprehensive enough to cope with any type of robot must be hand-coded and would work only for a single task, which would be unfeasible or very hard to maintain in the real world. Moreover, they only consider transferring value functions, hence further investigations could explore how to facilitate the ontology construction or how to transfer knowledge between agents following different learning algorithms (e.g., how to transfer knowledge between a *teacher* learning through Q-learning and a *student* applying policy search?).

Domains that require long wallclock time to execute actions are very hard to solve through RL. *HogRider* [Xiong et al., 2018] deals with a domain that requires several seconds to execute a single episode, posing a challenge to exploration. This problem is solved by building a tree specifying rules for initiating the *student*'s Q-table. Those rules are manually defined by a human and are used for providing a reasonable initial policy when starting learning. Although *HogRider* requires extensive domain knowledge, it achieves an impressive performance in this challenging setting. Therefore, this method represents a valuable effort toward adapting TL procedures to more complex and realistic tasks.

CHAPTER 6

Experiment Domains and Applications

Prospective domains for multiagent RL are the ones in which multiple agents are present in the environment and one or more of them are trying to solve a task that can be codified through a reward function. For all of those domains, one or more categories of TL might be applied. Therefore, a myriad of real-world and simulated domains could be used for validation and performance comparisons. In practice, some domains are preferred in the literature, where the following characteristics are desired.

1. Humans can learn to perform the task and estimate relative performances.

2. An interface is available (or could be developed) for human understanding and interaction.

3. The interval between interactions with the environment is adjustable or compatible with human reaction-times, so as to enable human participation.

4. Agents have the ability to observe each other and to communicate.

5. Taking into account the other agents in the environment might improve the performance of one agent, either by coordinating with teammates or by adapting against the strategy of opponents.

6. Multiple (or at least non-obvious) optimal policies exist for solving the task.

7. Multiple scenarios with varying degrees of difficulty and similarity exist or might be built for posteriorly reusing knowledge.

In the next subsections, we discuss the most prominent domains for evaluation of multiagent transfer methods.

6.1 GRIDWORLD AND VARIATIONS

The *Gridworld* domain has been extensively used for evaluations of both single and multiagent TL algorithms.

This domain is easy to implement, customize, and analyze results. Moreover, it is easy to develop simple interfaces that make the task to be solved very intuitive even for laypeople. For

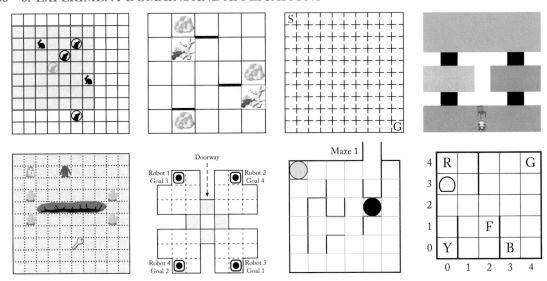

Figure 6.1: Illustrations of different settings of the *Gridworld* domain. From top-left to bottom-right, based on: *Predator–Prey* [Silva and Costa, 2017a], *Goldmine* [Diuk, 2009], *Multi-room Gridworld* [Kolter et al., 2008], *Dog Training* [Peng et al., 2016a], *Gridworld with beacons* [Narvekar et al., 2017], *Multiagent Gridworld Navigation* [Hu et al., 2015a], *Theseus and the Minotaur* [Madden and Howley, 2004], *Taxi* [Dietterich, 2000].

this reason, *Gridworld* has become a *first trial* domain for MAS, where methods are validated and tuned before applied in more complex problems.

Basically, any *Gridworld* domain has a group of agents that can navigate in an environment for solving a task.

While the most classic implementation is a simple navigation task in which the agents aim at reaching a desired destination, popular variations include *Predator–Prey*, where a group of "predator" agents aims at capturing "prey" agents, and *Goldmine*, in which a group of miners aims at collecting gold pieces spread in the environment. Figure 6.1 shows the diversity of settings that can be easily built with small variations in the *Gridworld* domain.

Despite being simple, results observed in this domain allow the evaluation of performance in terms of collaboration and adaptation (e.g., avoiding collisions or collaboratively capturing prey), robustness to noisy sensors and actuators (e.g., including noise in observations or a probability of incorrectly executing chosen actions), and scalability (e.g., how big can be the *Gridworld* that the technique solves?). It is also easy to create more complex versions of a task by trivially increasing the size of the grid or the number of objects in the environment.

Moreover, *Gridworld* tasks are very intuitive for humans and can be used for receiving knowledge from laypeople without much effort for explaining the task [Peng et al., 2016a].

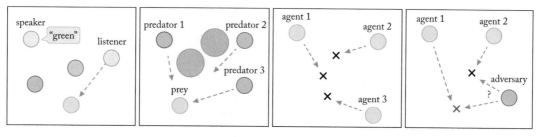

Figure 6.2: Illustration of different multiagent tasks that can be built from the *particle* environment (based on Lowe et al. [2017]).

Typically, *Gridworld* environments have discrete state and action spaces (agents move from one cell to another in the grid). However, the recent research trends required environments where function approximation is needed either because of a huge state space or continuous action set. As the *Gridworld* environment did not lose its appeal, the *particle* [Lowe et al., 2017] environment (a Deep RL version of *Gridworld*) is becoming increasingly popular.

Figure 6.2 illustrates the particle environment. The same high-level idea applies for this environment, but in this case the state and action sets are continuous. This domain is still easy to visualize and interpret, but hard enough that very naive tabular RL will not work.

In summary, *Gridworld* and its variations will always be an important asset for the development of TL algorithms.

6.2   SIMULATED ROBOT SOCCER

RoboCup [Kitano et al., 1997] was proposed as a long-term challenge for the AI community. Although playing soccer with real robots involves various implementation issues not always related to AI problems, *simulated* robot soccer has long been used as a testbed for multiagent RL algorithms.

A very simplified *Gridworld*-style soccer game was introduced by Littman [1994] for evaluating adversarial multiagent algorithms (Figure 6.3). More recently, the RoboCup's 2D simulation league [RoboCup, 2019] was shown to be an ideal testbed for multiagent TL techniques. In robot soccer tasks, strategies and moves might be imitated from teammates or opponents [Floyd et al., 2008], communication can be established to transfer knowledge, teleoperation is possible, and knowledge can be reused from previous games [Bianchi et al., 2009].

As deploying the full RoboCup simulation might still be hard, some simplifications of the robot soccer task have become popular. *KeepAway* [Stone et al., 2005] consists of learning how to maintain possession of the ball (Figure 6.3), and *Half Field Offense* [Hausknecht et al., 2016] provides a simplified goal-scoring task by using half of the field (Figure 6.3). Those are still hard learning problems. Agents have to cope with continuous observations, noisy sensors and actuators, and uncertainties in regard to other agents' policies. It is also easy to create harder

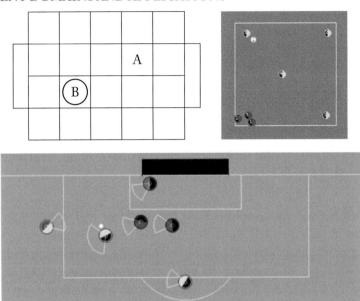

Figure 6.3: Illustrations of the different Robot Soccer simplificated simulations. From top-left to bottom: *Littman's Soccer* [Littman, 1994], *KeepAway* [Stone et al., 2005], and *Half Field Offense* [Hausknecht et al., 2016]. Used with permission.

(or easier) tasks by trivially manipulating the number of agents in the task. Knowledge reuse across different soccer environments is also possible [Kelly and Heywood, 2015].

6.3   VIDEO GAMES

Playing video games requires the ability to solve a task while creating and executing strategies to deal with other agents (that either learn or follow fixed strategies). For this reason, "old school" games have been used as challenging learning problems for RL methods. Moreover, those games present an ideal scenario for knowledge reuse, as it is common for games to present an easier task (level) and proceed to more complex ones as the player successfully solves the first.

Figure 6.4 depicts some of the games that have been used as validation domains. Usually, state features specially tailored for the game at hand are used as input to the learning agents. Those features are either extracted from an image processing procedure [Diuk et al., 2008] or through using internal variables of the game codification [Taylor et al., 2014a]. While the former is harder to compute for an automated agent, the latter usually forms state features that are meaningless for humans. Nevertheless, the literature presents a good number of successful approaches for solving various games.

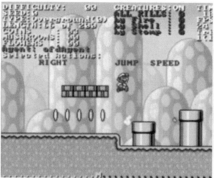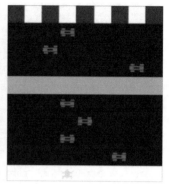

Figure 6.4: Some examples of games that have been used for evaluating TL techniques. From left to right: *Pacman* [Svetlik et al., 2017], *Mario* [Krening et al., 2017], and *Frogger* [Subramanian et al., 2016].

Most of the approaches so far transfer knowledge from other agents (especially humans). However, knowledge could be transferred from other levels or even games. Even when having very different objectives, games often have similarities (such as using the same buttons for moving the character). However, autonomously computing similarities and mappings between games is still an open problem [Glatt et al., 2016, Du et al., 2016].

After Deep RL has been introduced, solving tasks directly from pixels became orders of magnitude easier. Hence, achieving super-human performance in very challenging, relatively newer, games became a way to prove the effectiveness of learning methods. Games are especially good for doing that because everything happens in a virtual world (that can even be accelerated during training time) that can be executed many times in parallel. Moreover, humans are forced to act in this environment through keyboard and mouse or joystick commands and to sense the environment through screens. This makes simulating the human ability in playing games very easy. Some examples of recently solved games are Dota 2 [Berner et al., 2019] and Starcraft II [Vinyals et al., 2019], illustrated in Figure 6.5. Although progressively harder games are likely to be solved in the next years, it is usually a better idea to validate methods in simpler games unless a huge infrastructure is available for providing computational power.

6.4 ROBOTICS

Deploying autonomous robots in the real world is one of the ambitions of multiagent RL. However, physical robotic platforms have to cope with problems that are not directly modeled in MDPs (or SGs). Noise in sensors and actuators and limited computational resources are some of the many challenges that hamper the direct application of RL methods in robots. Moreover, random exploration is prone to destroy the robot or even harm people or animals in the environment, thus TL is required for successfully deploying a physical robot using RL.

Figure 6.5: Screenshots from Dota 2 (left) and Starcraft II (right), arguably the most complex games yet solved using learning approaches.

In spite of those challenges, RL methods allied with TL techniques have been successful in solving small tasks, usually by interacting with humans. Some examples of successful applications of TL in the literature are *Robotic Arm Control* [Schaal, 1997], *Robotic Path Planning* [Kolter et al., 2008], *Humanoid Robot Control* [Sakato et al., 2014], and *Quadrimotor Control* [Isele et al., 2016].

As the methods proposed in the literature are still not able to easily handle tasks in which several robots are present, this domain was mainly used to evaluate *Inter-Agent* transfer methods with a single robot. While those small robotic tasks will continue to be explored and refined, the reuse of knowledge from simulators for guiding the exploration in the real world is an especially promising application to be explored [Hanna and Stone, 2017].

MuJoCo [Todorov et al., 2012] is a popular framework for building simulated robotic-like control tasks. Figure 6.6 illustrates some of the most common MuJoCo tasks for RL research. The learning agent has to learn how to move a particular body through applying torque in its joints. *Ant* represents an easy task to solve, while *Humanoid* is harder because of the body complexity. This framework is especially popular for evaluating some *Inter-Agent* transfer approaches [Pinto et al., 2017, Gupta et al., 2017a, Devin et al., 2017].

6.5   SMART GRID

Taylor et al. [2014b] proposed a solution for one of the most realistic applications of TL in RL. As illustrated in Figure 6.7, multiple electric vehicles are connected to a single transformer (forming a neighborhood), and they all need to recharge their batteries for completing daily travels. However, as many people work on similar shifts, many cars are recharging during the peak hours, causing undesirable spikes on energy demands.

The goal of the agents (vehicles) is to coordinate their charging times for avoiding peak hours and alternate themselves when charging, so as to keep the transformer load always at acceptable levels.

Figure 6.6: Examples of MuJoCo control tasks. From left to right: *Ant*, *Cheeta*, and *Humanoid*. Most commonly, the learning agent has to learn how to move the body as quickly as possible. As the body becomes more complex, it becomes harder to learn how to coordinate joints through random exploration.

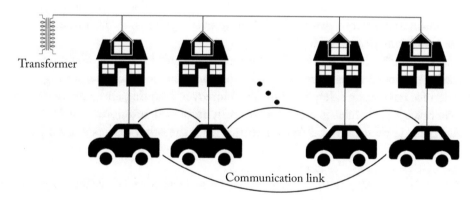

Figure 6.7: Illustration of a common smart grid domain. A neighborhood of houses is connected to a transformer, where each of the houses has an electric vehicle. The cars need to recharge their batteries, but many of them recharging at the same time might cause transformer overloads.

Many knowledge reuse opportunities remain unexplored for this domain. A newcomer car could receive knowledge from the others in the neighborhood for accelerating adaptation and learning or policies could be reused in another neighborhood with similar characteristics, for example. As most of real applications, many technical challenges are yet to be solved, such as how to handle transfer of knowledge between cars produced by different manufacturers.

Figure 6.8: Visual output of a possible SMARTS environment.

6.6 AUTONOMOUS DRIVING SIMULATION

Building autonomous driving cars that can be used routinely is an ambition that attracted investment from many companies and stimulated the imaginary of scientists and laypeople alike. In practice, autonomous cars already exist but have been applied only in limited locations and in favorable conditions [Badue et al., 2021].

The lack of social ability is one of the current challenges for scaling up autonomous driving. In heavily trafficked areas, certain social behaviors emerge, and interpreting those social cues is needed for navigating safely and quickly. Moreover, adapting and quickly reacting when other drivers break transit rules is crucial. Currently, employed autonomous vehicles are overly conservative because they are not able to interpret what the other drivers will do [Zhou et al., 2020]. In its turn, this conservative behavior might cause unnecessary congestion or frustration to the human drivers behind the autonomous car.

While multiagent RL could enable those social and coordination abilities, needless to say, testing algorithms in autonomous vehicles is prohibitively expensive and might be dangerous. For this reason, the *SMARTS* environment was developed for providing a simulated autonomous driving environment where behavior models can be learned and evaluated.

SMARTS, illustrated in Figure 6.8 and available at Zhou et al. [2020], is a promising simulated environment for learning and evaluating coordination strategies. *SMARTS* is natively multiagent and enables building experiments in multi-process cloud-based platforms. Further investigations can leverage the platform to evaluate transfer methods in a complex domain, and perhaps to further advance the state-of-the-art in self-driving vehicles making use of TL and multiagent RL.

CHAPTER 7

Current Challenges

As shown in the previous chapters, TL for MAS has already played an important role toward scaling RL to more complex problems. However, many research questions are still open. In this section, we discuss prominent lines of research that will require further investigations in the years to come.

7.1 CURRICULUM LEARNING IN MULTIAGENT SYSTEMS

The ideas introduced by *curriculum* learning techniques can greatly benefit RL agents in MAS. However, the state-of-the-art still needs to be significantly improved in order to be useful in complex tasks. A potential use of *curriculum* learning in MAS could be, for example, training how to play a game against simulated opponents who become progressively more competent, and then reuse the gathered knowledge to face a real-world highly specialized opponent. Self-play for improving the policy has been extensively used before (a notable example is AlphaGo by Silver et al. [2016]), but games are usually played arbitrarily while *curricula* are expected to be much more efficient and only present useful tasks for the agent.

The *Transfer of curriculum*, as illustrated in Figure 7.1, is also an open and relevant problem (which has as early techniques the ones proposed in Section 5.7). We expect that future methods will be able to adapt an existent *curriculum* to a different agent, tailoring it according to the agent's unique capabilities and particularities. This is a very challenging goal, as the information about the target task is usually scarce for RL agents, which is further complicated in MAS where the strategy of other agents is unknown.

Indeed, Leibo et al. [2019] argue that the "intelligence" that can be extracted from a single-agent environment is bounded. On the other hand, whenever the system is composed of multiple adaptive units, a myriad of learning problems naturally emerges from the interaction between agents and changes in strategies (what they call as *Autocurricula*). Therefore, there are many avenues of possible developments and investigations in the area of *curriculum* learning for multiagent scenarios.

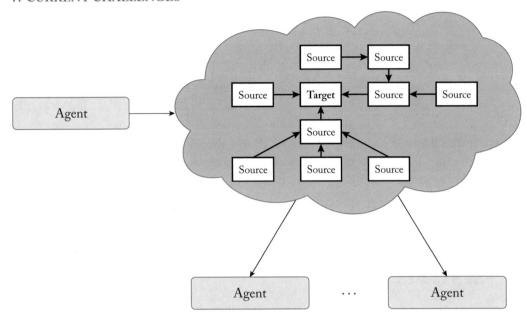

Figure 7.1: Illustration of a possible *curriculum* transfer scenario. After one agent learns a *curriculum*, it can be used and possibly refined by other agents.

7.2   BENCHMARKS FOR TRANSFER IN MULTIAGENT SYSTEMS

Numerically comparing two algorithms is very useful for the AI community, as a proper metric is an impartial and clear method for analyzing the relative performance when solving a given problem. However, no single TL metric is sufficient to have a clear picture of the performance, and even all of the currently used ones together might be inefficient for MAS techniques.

For understanding how evaluations of TL methods might be misleading, consider the fictitious performances in Figure 7.2a. *Alg1* seems to have a better performance in this graph, as both jumpstart and end of learning performances are higher than the one observed for *Alg2*. Now suppose that we train the agents for longer and the result is observed in Figure 7.2b. Notice that now *Alg2* seems to be better, because in spite of the lower jumpstart the end of learning performance is much higher. Therefore, choosing among *Alg1* and *Alg2* is simply arbitrary if the number of learning episodes is chosen without care. This analysis is further complicated if we consider costs of communication for transfer of knowledge between agents. If one method has a better performance but transfers a much higher number of messages, which of them can be considered the best one? There is no single answer to this question, and performance comparisons must be carried out in a very domain-specific and subjective way. The difficulty of comparing

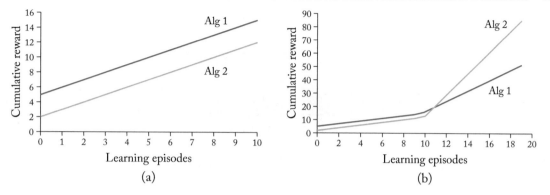

Figure 7.2: Fictitious performance comparison of two algorithms. (a) Running 10 episodes; (b) running 20 episodes.

methods and the differences in assumptions across TL methods are partially responsible for the lack of comparisons between methods in a significant portion of the literature.

Hyperparameter optimization is also an issue when comparing RL algorithms in general. Since the performance of most algorithms is very sensitive to multiple parameters, a designer bias is introduced in the evaluation. Then, it becomes hard to define whether a performance improvements comes from using a better algorithm or just from selecting better parameters. Integrating the cost of hyperparameter search in the evaluation is an option to make the evaluation fair [Jordan et al., 2020]. However, this would increase significantly the number of repetitions needed for extracting performance metrics.

Therefore, one of the most important open problems in the TL for MAS area is the development of proper comparison methods. In addition to the introduction of new metrics, the definition of standard benchmarks for comparing algorithms is required, similarly as proposed by Vamplew et al. [2011] for *Multiobjective* RL.

Benchmark problems would allow quickly deploying a new TL algorithm and comparing it to similar previous methods. Moreover, evaluating algorithms across a wide range of domain and scenarios will ensure that the algorithm is not severely overfitted to a single problem. Notice that the characteristics listed in Chapter 6 are the ones desired for domains used to evaluate multiagent TL algorithms. Although the domains listed in the previous chapter could all be part of the set of recommended benchmark domains, each paper tends to perform its own parameterization, to choose arbitrarily the length of the training process, and to pick its own evaluation metrics. Ideally, there should be benchmark directives stating clearly what each domain evaluates, and how to configure the domain and experiment for that purpose.

We hope that the taxonomy proposed here, as well as our suggestions of desired domain characteristics, will help the community to propose benchmarks for each category of TL methods discussed in this book.

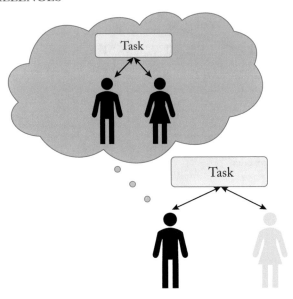

Figure 7.3: Illustration of a scenario in which the agent seeks to reuse previously acquired knowledge in a new task. However, the source task had different teammates, which means that the agent has to adapt itself to a team possibly following a different strategy.

7.3   KNOWLEDGE REUSE FOR AD HOC TEAMS

In *Ad Hoc* teams [Stone et al., 2010], agents have to solve collaborative tasks with previously unknown teammates, which means that no predefined communication protocol is available when starting learning. For this reason, learning to coordinate with each new teammate takes some time, and the exchange of information is often unfeasible due to the lack of a common protocol.

Although there has been developments on autonomously learning communication protocols [Foerster et al., 2016], those approaches have to solve a hard learning problem before any language emerges. Due to the dynamic nature of *Ad Hoc* tasks, the team of agents is likely to have changed before any language emerges. Therefore, learning how to communicate is unfeasible unless the task to solve is so hard that it will take much longer to solve it than to learn a language. However, it might be possible to reuse knowledge gathered from playing with another similar teammate, even if the agents are not able to communicate. For this purpose, the learning agent would need to be able to estimate the similarity between the current and past teammates, and also to properly reuse the previous knowledge (see Figure 7.3).

Performing knowledge reuse would be especially challenging if teammates do not follow a fixed strategy and are also learning. In this case, computing similarities and reusing knowledge could be harder, but would also make the RL agent more flexible and robust to sudden changes in the system.

7.4 END-TO-END MULTIAGENT TRANSFER FRAMEWORKS

Although usually enabling the RL agent to learn autonomously after deployed in the environment, the current state-of-the-art methods still require considerable human effort and expertise in order to be applicable. Most of the methods require parameters that must be specified in a domain-specific manner, and communication and knowledge transfer protocols are usually implicitly hard-coded in the agent's implementation.

All those definitions, that are painstakingly optimized for each domain, hurt the applicability of the methods, and for this reason, most of the successful applications of the methods here discussed are in controlled environments.

In order to create more robust agents, the community needs to create a new line of research on the development of end-to-end transfer learners. The main goal of this line would be building agents able to autonomously create, optimize, and use protocols for negotiating knowledge with other agents, tune their parameters, and find the best way to combine internal knowledge with received information. The initial works on autonomously learned communication [Foerster et al., 2016, Sukhbaatar et al., 2016] are related to this challenge and might be seen as a first step toward autonomously learning transfer protocols.

Specifically, this line of research would enable a methodological answer to the general question "who should initiate the transfer?" While action advising approaches (Section 5.1) have been a little more aware and explicit about this challenge, the burden of the decision of exactly when the transfer should happen has historically been placed arbitrarily in *Inter-Agent* TL methods. Ideally, both agents involved in the TL relation should be able to initiate transfer when they are able to assess an opportunity.

7.5 TRANSFER FOR DEEP MULTIAGENT REINFORCEMENT LEARNING

Deep RL has achieved many impressive successes in recent years [Silver et al., 2016], especially for problems where it is hard to extract useful engineered features from raw sensor readings (such as images) and is now starting to be employed in MAS [Castaneda, 2016, Gupta et al., 2017b]. Even though all the ideas presented here could be used to some extent for Deep RL, most of the methods are not directly usable in agents learning with deep networks. Therefore, the adaptation of transfer methods for deep multiagent RL is a promising area of research.

The preliminary investigations discussed in Sections 4.7 and 5.8 showed that Deep RL agents also benefit from reuse of knowledge, as shown in Figure 7.4, but the effects of negative transfer could be catastrophic if unprincipled transfer is performed [Glatt et al., 2016, Du et al., 2016]. Those initial efforts in this topic still need further investigations for scaling Deep RL to complex MAS applications.

Figure 7.4: Deep RL agents also benefit from the reuse of knowledge stored in deep neural networks. However, those agents have to be especially careful to avoid negative transfer, due to the difficulty of interpreting neural networks.

7.6   INTEGRATED INTER-AGENT AND INTRA-AGENT TRANSFER

As shown in the previous chapters, *Inter-* and *Intra-Agent* transfer are very rarely integrated, and most of the methods focus on one of them alone. However, humans routinely combine both types of knowledge reuse, and it is most likely that agents integrated into our domestic and corporate environments will have to consistently implement both types of knowledge reuse.

This challenge was already highlighted before [Silva and Costa, 2017b], and implemented to a very limited extent by some methods [Madden and Howley, 2004, Wang et al., 2016]. However, many problems inherent from combining knowledge from different sources remain unsolved. Imagine that a teacher is willing to provide demonstrations to the student, while a mentor is ready to be observed, and a previously learned policy from a similar problem is available. Should the agent follow the demonstration, the previous knowledge, or try to imitate the mentor? Would it be possible to combine knowledge from all those sources? Those and many other questions that would be unveiled by a deeper investigation are not answered by the current literature.

7.7   HUMAN-FOCUSED MULTIAGENT TRANSFER LEARNING

At first sight, transferring knowledge from a human seems to be easy. Humans are very good at abstracting knowledge and carry years of previous experiences from which automated agents could profit. However, humans have a different set of sensors, reaction times, and have a limited capability of focusing on a task. Those differences might lead the human to provide useless or even inconsistent knowledge in the point of view of the agent.

These problems are aggravated when knowledge is transferred from non-expert humans, which might not understand that automated agents follow a different reasoning process and do not have the large amount of previous knowledge and common-sense reasoning that humans have. Still, as AI experts represent a tiny portion of the human population, laypeople will be required to instruct automated agents in the near future. Therefore, even though already explored

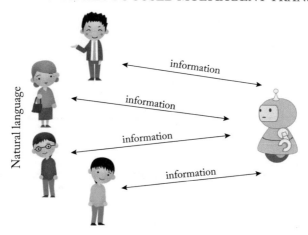

Figure 7.5: In scenarios where agents and humans coexist and exchange information, appropriate methods are needed to translate human natural language into a more appropriate language for artificial RL agents, and vice versa.

by a good portion of the *Inter-Agent* transfer literature, there is still much to be improved on human-focused transfer.

As a way to facilitate the participation of non-experts on agent training, additional investigations are needed to develop appropriate methods for translating common human languages into instructions that are useful for RL agents, as illustrated in Figure 7.5. Also, it is desired that the methods take advantage of the psychological bias to incentive humans to give good instructions, for example, reducing the agent's actuation speed to indicate that the agent has a high uncertainty for a given state, as Peng et al. [2016a] do.

The current literature also neglects issues related to human reaction times. Simulated environments are often paused for receiving human feedback, which is impossible for many tasks in the real world. This is especially important for robotic tasks, as a delayed corrective feedback could lead to a harmful movement that cannot be interrupted in time for avoiding a collision.

Transfer methods are also desired to be easily readable and understandable for humans [Ramakrishnan et al., 2016], as "black-box" methods should not be employed in critical applications.

Although extracting consistent knowledge from humans might be hard, they often have a privileged point of view in the task. This is due to either because they have modeled the task themselves or because of their vast previous knowledge, which are the reason why humans have been a popular source of knowledge in TL approaches after all. Therefore, there is still room for improvements on how humans can share their knowledge with the agents. One low-hanging idea to explore would be asking humans to estimate "how much" of the received reward was due to the agent individual actuation instead of the teammates actuation. Figuring this out is very

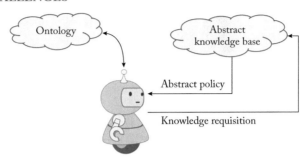

Figure 7.6: Illustration of how a web-based knowledge base could be built. The agent accesses an ontology base for defining abstract state and action spaces. Then, an online knowledge base provides knowledge to the agent.

difficult and known as a credit assignment problem [Devlin et al., 2014]. This issue is pervasive in multiagent learning approaches.

7.8   CLOUD KNOWLEDGE BASES

Kono et al. [2014] introduced the innovative idea of having an ontology available in the web for translating the agent's individual capabilities into a more abstract set of skills, which would enable the transfer of knowledge between two robots manufactured by different companies, for example.

Their idea could be extended for building *web–based knowledge bases*, that could be used by agents as a common knowledge repository to extract and add knowledge (playing the role of the internet in the modern human life). The idea is illustrated in Figure 7.6.

When a newly built agent is joining a system or starts learning in a task, it will make use of an ontology to translate its own sensor readings and low-level actions into abstract states and skills that are common to all similar robots. Then, an abstract policy could be used to bootstrap learning, or alternatively, the base could inform the IPs of similar robots with which the agent could share knowledge.

The major challenge of implementing such architecture would be preventing agents from adding incorrect or malicious knowledge into the knowledge base, which is related to the challenge in Section 7.10.

7.9   MEAN-FIELD KNOWLEDGE REUSE

Scaling multiagent RL to problems with many agents is one of the major ambitions of the area. Yet, as shown throughout this book, most of the recent developments apply (or at least has been only evaluated in) for domains with few agents.

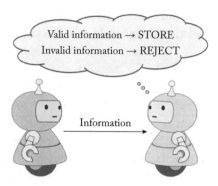

Figure 7.7: Ideally, RL agents should be able to verify the validity of any information received. However, this is very challenging in practice.

Mean-Field RL [Yang et al., 2018] factorizes the value function based on interactions between pairs of agents. This means that a Q function has to be learned only considering the power set of two action sets, instead of considering all the joint action space. Hence, the method is scalable for a very high number of agents.

However, the main limitation of the method is that those agents have to be homogeneous. There are initiatives toward relaxing this assumptions [Subramanian et al., 2020], but the best effort today relies on a manual separation of the agents in groups of homogeneous agents, which does not quite solve the problem.

Therefore, embedding knowledge reuse in mean-field-inspired RL approach could be a promising avenue for applying RL in domains with many agents. Adaptations of many TL methods could be used to accelerate learning, or to ease adaptation across different clusters of agents using Mean-Field RL.

7.10   SECURITY

Nearly all of the current literature assumes that agents involved in knowledge reuse interactions are benevolent, i.e., they will never communicate wrong or inconsistent knowledge on purpose, but this is rarely the case for real-world applications, as shown in Figure 7.7. In fact, in many of the works discussed here, the agent transferring knowledge is assumed to have perfect knowledge of the task and the optimal policy, while a more realistic assumption would be that a reasonable policy can be estimated through TL but the agent still needs to explore the environment for learning the optimal policy.

As MAS are progressively scaled for real-world applications, RL researchers need also to be concerned about security. If agents base their exploration on knowledge received from communication, malicious agents could be possibly able to transfer inconsistent or incorrect knowledge for affecting the learning agent's performance. For robotic systems or autonomous cars, a malicious communication could cause serious damages to very expensive equipments. If the agent is deployed in uncontrolled environments, it is also possible to cause damage to people, other agents, or buildings.

Therefore, the development of transfer methods that are robust to interference is imperative. Reward shaping methods (Section 5.5) have a solid theoretical background and are one of the few families of transfer algorithms proved to converge to the optimal policy even when malicious transfer is performed. For this reason, those methods might have an important role toward applications in which security is critical. Another possible step toward secure methods could be the development of argumentation protocols, in which agents will have to provide arguments explaining why the transferred knowledge is reliable and correct. Deploying security into knowledge-transferring agents is a very challenging but necessary research goal.

7.11   INVERSE REINFORCEMENT LEARNING FOR ENFORCING COOPERATION

Although IRL is most commonly used only for defining a reward function through demonstrations to learning agents, the recent trend on using such techniques for multiagent tasks opens up new research opportunities [Lin et al., 2018].

Natarajan et al. [2010] solve a multiagent cooperative problem by using IRL for learning the reward function of all the agents in the environment (which have a locally optimal reward). Then, a centralizer agent enforces the agents to change their actions to the expected global optimal policy, where they were not doing it already. Even though having a centralized agent dictating the joint actions is unfeasible to most domains, this agent could oversee the actuation of the other agents and introduce additional rewards as an incentive to cooperation. Such framework, if successfully developed, could be a possible solution for domains in which an institution wants to incentivize cooperation between self-interested agents (such as some of the techniques studied by the area of *Organization of MAS* [Argente et al., 2006]).

7.12   ADVERSARY-AWARE LEARNING APPROACHES

In a near future, every household or office will have a number of learning-based devices routinely performing tasks. Once we make that transition, learning-based agents will be subject to all kinds of tampering and hacking attempts (see Figure 7.8), maybe even including RL-specialized ones.

Therefore, we as a community need to understand how vulnerable are multiagent RL algorithms to adversarial tampering, both practically and theoretically. Tangkaratt et al. [2020b] contribute an initial work on understanding the theory behind designing an optimal environment tampering strategy to affect a learning agent's performance.

Additional investigations along this direction are needed so that we can understand (i) what are the best ways to tamper with or hack an RL system and (ii) what are the learning algorithms most subjected to interferences. Only through acquiring a deep understanding

Figure 7.8: Whenever there is an interface with the outside world (including communication with other agents), there will also be a window for knowledge tampering. Ideally, RL agents who communicate with each other should be robust to that.

on all possible attack strategies, will we be able to design learning algorithms and multiagent systems that are very resilient against malicious interferences.

This line of work is related to *Security* described in Section 7.10. However, here we must be resistant against attacks designed specially for RL agents.

CHAPTER 8

Resources

This chapter provides valuable pointers for interested researchers. We discuss the *conferences* and *journals* that most frequently publish TL for MAS papers, which will help both for searching for papers and for submitting new proposals. We also list code libraries that can help in the implementation of algorithms.

8.1   CONFERENCES

So far, no major conference is entirely devoted to TL for MAS. The *International Conference on Autonomous Agents and Multiagent Systems* (AAMAS) is the conference that fits best with the topic. However, comprehensive AI and Machine Learning conferences also have embraced papers on the topic over the past years.

Top-tier general AI conferences such as *International Joint Conference on Artificial Intelligence* (IJCAI), *AAAI Conference on Artificial Intelligence* (AAAI), *International Conference on Machine Learning* (ICML), *European Conference on Artificial Intelligence* (ECAI), and *European Conference on Machine Learning* (ECML) are of interest.

Other conferences from related topics often publish papers on TL for RL, where *Neural Information Processing Systems* (NeurIPS) and *International Joint Conference on Neural Networks* (IJCNN) must be highlighted. Finally, some of the Robotics conferences sometimes publish papers on the topic (usually with focus on the application), such as the *International Conference on Robotics and Automation* (ICRA), the *International Conference on Intelligent Robotics and Systems* (IROS), and the recent *International Symposium on Multi-Robot and Multi-Agent Systems* (MRS).

Workshops of major conferences are also known for publishing very inspiring ideas, even though they are not fully developed sometimes. The annual *Adaptive Learning Agents* Workshop (ALA) at AAMAS consistently publishes novel ideias on the topic. The *Scaling-Up Reinforcement Learning* (SURL) at ECML and IJCAI has also became a recurrent workshop focused on this topic. Every major conference periodically holds workshops on TL, such as the *Transfer in Reinforcement Learning* (TiRL) at AAMAS and the *Lifelong Learning Workshop* at ICML workshops.

8.2   JOURNALS

As no journal specializes in TL for MAS, most of the articles are submitted to a general AI/Machine Learning or an application-oriented journal. The most adherent journal would perhaps be the *Journal of Autonomous Agents and Multiagent Systems* (JAAMAS). However, relevant publications are also found in most of the high-impact general AI and Machine Learning journals such as the *Journal of Artificial Intelligence Research* (JAIR), *Journal of Machine Learning Research* (JMLR), *Artificial Intelligence, Machine Learning*, and *IEEE Transactions on Cybernetics*. The newly created journal *Machine Learning and Knowledge Extraction* (MAKE) is also a promising venue.

8.3   LIBRARIES

Implementing domains and baseline learning algorithms from scratch is common for researchers in the area, which hampers method comparisons. No multiagent RL framework is widely adapted by the community as a whole, but some libraries were publicly released for facilitating research in the area. BURLAP [MacGlashan, 2015] is an early example and implements some simple multiagent RL domains and algorithms (called as *Stochastic Games* domains) in the *Java* programming language.

However, due to the recent popularity of tensor-based *Python* frameworks such as *Tensorflow* [Abadi et al., 2015] and *PyTorch* [Paszke et al., 2019], virtually all of recent publications prefer Python-compatible RL frameworks.

Indeed, tensor-based computing facilitated implementing deep learning value approximators and porting code to different hardware platforms such as GPUs or TPUs. This ability is of utmost importance now that any innovative application relies on cloud computing and/or dedicated hardware platforms [Vinyals et al., 2019, Berner et al., 2019].

In spite of not supporting the development of multiagent environment, RL-Glue [Tanner and White, 2009] has been used to validate *Inter-Agent* transfer applied to single-agent problems. It consists of a multi-language library for providing a standard repository for RL algorithms independent on the preferred programming language.

In recent years, *OpenAI Gym* [Brockman et al.] emerged as a promising framework for RL agents. In the current version, agents can be easily built in *Python*. In spite of not focusing on MAS, some extensions for this purpose have been contributed by the community.

RLlib [Liang et al., 2018] provides implementations of multiagent RL algorithms and environments. This framework facilitates scaling up the implementations to clusters and containerized infrastructures, therefore it is a promising framework for practitioners in the area, especially in practical applications.

Notice that none of those libraries are specialized for TL, which hampers the development of the area because of the difficulty of comparing similar approaches in the literature. The development of modules within the already-existent libraries could assist in both the development of the literature and the availability of TL methods for non-experts.

CHAPTER 9

Conclusion

Artificial Intelligence has been used to solve challenging real-world tasks where learning and adaptation abilities are required for delivering reasonable solutions. Reinforcement Learning (RL) is perhaps the most effective learning paradigm for sequential decision-making problems. RL has rapidly reached new applications in recent years in a pace that promises many breakthrough developments in the next few years.

As the tasks we need to solve get more complex, centralized and single-agent solutions become infeasible. Therefore, building Multiagent Systems (MAS) is a natural step, since most of human achievements came from collaborative learning and developments. While many RL techniques can be adapted to multiagent scenarios, new particular challenges arise, such as limitations in communication, the explosion of the state-action space, and the value assignment problem across agents.

Most relevant multiagent RL tasks are complex enough that randomly exploring the environment is infeasible. Therefore, additional techniques are needed for accelerating learning. Knowledge reuse has gone long ways for multiagent RL agents, enabling the solution of tasks unsolved otherwise. However, the current literature is divided into groups of methods that have little interoperability, which makes it difficult to build an agent that benefits from all types of knowledge reuse.

This book described a taxonomy for TL in MAS, aiming at both helping newcomers in the area to have an overview of the current literature and at incentivizing seasoned researchers to observe similarities and research opportunities across different current lines of research. Recent approaches are grouped into *Intra-* and *Inter-Agent* transfer methods, which correspond to knowledge reuse from experience gathered from the own agent and from other agents, respectively. Equipped with our categorization, we hope that the interested reader will be able to assimilate and to distinguish the different nuances, pros and cons, and capabilities of all groups of methods. Most importantly, we expect that the reader will be able to easily identify which of the groups is appropriated to solve her or his problem, and what are the crossing points with other approaches.

Although by no means exhausting all works in the area, we also contribute in-depth discussions of representative works showing similarities, deficiencies, and opportunities for future developments. Our discussions might be useful for newcomers in the area that are in early stages of their research and looking for an specific problem to work on. As for seasoned researchers, we hope that our discussions will enable the identification of crossing points with other communities, sparking new collaborations.

In conclusion, we hope that this book will encourage the community work together to solve the challenging problem of reusing knowledge from multiple sources in an autonomous and secure manner, difficult, yet necessary goal for bridging the gap between our current literature and complex real-world multiagent RL applications.

Bibliography

Richard S. Sutton and Andrew G. Barto. *Reinforcement Learning: An Introduction*, 2nd ed., MIT Press, Cambridge, MA, 2018. 1, 7

Michael L. Littman. Reinforcement learning improves behaviour from evaluative feedback. *Nature*, 521(7553):445–451, 2015. DOI: 10.1038/nature14540 1

Gerald Tesauro. Temporal difference learning and TD-gammon. *Commun. ACM*, 38(3):58–68, 1995. DOI: 10.1145/203330.203343 1

David Silver, Aja Huang, Christopher J. Maddison, Arthur Guez, Laurent Sifre, George van den Driessche, Julian Schrittwieser, Ioannis Antonoglou, Veda Panneershelvam, Marc Lanctot, Sander Dieleman, Dominik Grewe, John Nham, Nal Kalchbrenner, Ilya Sutskever, Timothy Lillicrap, Madeleine Leach, Koray Kavukcuoglu, Thore Graepel, and Demis Hassabis. Mastering the game of Go with deep neural networks and tree search. *Nature*, 529:484–503, 2016. DOI: 10.1038/nature16961 1, 73, 77

Oriol Vinyals, Igor Babuschkin, Wojciech M. Czarnecki, Michaël Mathieu, Andrew Dudzik, Junyoung Chung, David H. Choi, Richard Powell, Timo Ewalds, Petko Georgiev, et al. Grandmaster level in StarCraft II using multi-agent reinforcement learning. *Nature*, 575(7782):350–354, 2019. DOI: 10.1038/s41586-019-1724-z 1, 34, 41, 69, 86

Susan M. Shortreed, Eric Laber, Daniel J. Lizotte, T. Scott Stroup, Joelle Pineau, and Susan A. Murphy. Informing sequential clinical decision-making through reinforcement learning: An empirical study. *Machine learning*, 84(1-2):109–136, 2011. DOI: 10.1007/s10994-010-5229-0 1

Jens Kober, J. Andrew Bagnell, and Jan Peters. Reinforcement learning in robotics: A survey. *The International Journal of Robotics Research*, 32(11):1238–1274, 2013. DOI: 10.1177/0278364913495721 1

Andrew G. Barto, P. S. Thomas, and Richard S. Sutton. Some recent applications of reinforcement learning. In *Proc. of the 18th Yale Workshop on Adaptive and Learning Systems*, 2017. 1

Ana L. C. Bazzan. Beyond reinforcement learning and local view in multiagent systems. *Künstliche Intelligenz*, 28(3):179–189, 2014. DOI: 10.1007/s13218-014-0312-5 1

Matthew E. Taylor and Peter Stone. Transfer learning for reinforcement learning domains: A survey. *Journal of Machine Learning Research (JMLR)*, 10:1633–1685, 2009. DOI: 10.1145/1577069.1755839 1, 2, 3, 19, 25

Matthew E. Taylor, Peter Stone, and Yaxin Liu. Transfer learning via inter-task mappings for temporal difference learning. *Journal of Machine Learning Research (JMLR)*, 8(1):2125–2167, 2007. 1

Felipe Leno Da Silva, Ruben Glatt, and Anna Helena Reali Costa. Simultaneously learning and advising in multiagent reinforcement learning. In *Proc. of the 16th International Conference on Autonomous Agents and Multiagent Systems (AAMAS)*, pages 1100–1108, 2017. 1, 46, 48

David Isele, Mohammad Rostami, and Eric Eaton. Using task features for zero-shot knowledge transfer in Lifelong Learning. In *Proc. of the 25th International Joint Conference on Artificial Intelligence (IJCAI)*, pages 1620–1626, 2016. 1, 25, 34, 39, 70

Jivko Sinapov, Sanmit Narvekar, Matteo Leonetti, and Peter Stone. Learning inter-task transferability in the absence of target task samples. In *Proc. of the 14th International Conference on Autonomous Agents and Multiagent Systems (AAMAS)*, pages 725–733, 2015. 1, 34, 39

Felipe Leno Da Silva, Garrett Warnell, Anna Helena Reali Costa, and Peter Stone. Agents teaching agents: A survey on inter-agent transfer learning. *Autonomous Agents and Multiagent Systems*, 34(9):2020a. DOI: 10.1007/s10458-019-09430-0 2

Alessandro Lazaric. *Transfer in reinforcement learning: A framework and a survey*, pages 143–173. Springer Berlin Heidelberg, Berlin, Heidelberg, 2012. DOI: 10.1007/978-3-642-27645-3_5 2, 3, 19

Adam Bignold, Francisco Cruz, Matthew E. Taylor, Tim Brys, Richard Dazeley, Peter Vamplew, and Cameron Foale. A conceptual framework for externally-influenced agents: An assisted reinforcement learning review. *ArXiv Preprint ArXiv:2007.01544*, 2020. 2, 3

David E. Goldberg. *Genetic Algorithms in Search, Optimization and Machine Learning*, 1st ed., Addison-Wesley Longman Publishing Co., Inc., 1989. 3

Lucian Busoniu, Robert Babuska, and Bart De Schutter. A comprehensive survey of multi-agent reinforcement learning. *IEEE Transactions on Systems, Man, and Cybernetics, Part C: Applications and Reviews*, 38(2):156–172, 2008. DOI: 10.1109/tsmcc.2007.913919 3, 15

Peter Stone and Manuela Veloso. Multiagent systems: A survey from a machine learning perspective. *Autonomous Robots*, 8(3):345–383, 2000. DOI: 10.21236/ada333248 3

Pablo Hernandez-Leal, Bilal Kartal, and Matthew E. Taylor. A survey and critique of multi-agent deep reinforcement learning. *Autonomous Agents and Multiagent Systems*, 33:750–797, 2019. DOI: 10.1007/s10458-019-09421-1 3

Thanh Thi Nguyen, Ngoc Duy Nguyen, and Saeid Nahavandi. Deep reinforcement learning for multiagent systems: A review of challenges, solutions, and applications. *IEEE Transactions on Cybernetics*, 50(9):3826–3839, 2020. DOI: 10.1109/tcyb.2020.2977374 3

Brenna D. Argall, Sonia Chernova, Manuela Veloso, and Brett Browning. A survey of robot learning from demonstration. *Robotics and Autonomous Systems*, 57(5):469–483, 2009. DOI: 10.1016/j.robot.2008.10.024 3, 26

Shao Zhifei and Er Meng Joo. A survey of inverse reinforcement learning techniques. *International Journal of Intelligent Computing and Cybernetics*, 5(3):293–311, 2012. DOI: 10.1108/17563781211255862 3, 26, 58

Felipe Leno Da Silva and Anna Helena Reali Costa. A survey on transfer learning for multiagent reinforcement learning systems. *Journal of Artificial Intelligence Research (JAIR)*, 64:645–703, 2019. DOI: 10.1613/jair.1.11396 3

Michael J. Wooldridge. *An Introduction to MultiAgent Systems*, 2nd ed., Wiley, 2009. 5

Paul Bogg, Ghassan Beydoun, and Graham Low. When to use a multi-agent system? In The Duy Bui, Tuong Vinh Ho, and Quang Thuy Ha, Eds., *Intelligent Agents and Multi-Agent Systems*, pages 98–108, Springer Berlin Heidelberg, Berlin, Heidelberg, 2008. DOI: 10.1007/978-3-540-89674-6_13 5

Martin L. Puterman. *Markov Decision Processes: Discrete Stochastic Dynamic Programming*. John Wiley & Sons, Hoboken, NJ, 2005. DOI: 10.2307/2291177 7

Christopher J. Watkins and Peter Dayan. Q-learning. *Machine Learning*, 8(3):279–292, 1992. DOI: 10.1007/bf00992698 9

Ian J. Goodfellow, Yoshua Bengio, and Aaron Courville. *Deep Learning*. MIT Press, Cambridge, MA, 2016. http://www.deeplearningbook.org 11

Volodymyr Mnih, Koray Kavukcuoglu, David Silver, Andrei A. Rusu, Joel Veness, Marc G. Bellemare, Alex Graves, Martin Riedmiller, Andreas K. Fidjeland, Georg Ostrovski, Stig Petersen, Charles Beattie, Amir Sadik, Ioannis Antonoglou, Helen King, Dharshan Kumaran, Daan Wierstra, Shane Legg, and Demis Hassabis. Human-level control through deep reinforcement learning. *Nature*, 518(7540):529–533, 2015. DOI: 10.1038/nature14236 11, 12

Matthew Hausknecht and Peter Stone. Deep recurrent Q-learning for partially observable MDPs. In *AAAI Fall Symposium on Sequential Decision Making for Intelligent Agents (SD-MIA15)*, November 2015. 12

John Schulman, Sergey Levine, Philipp Moritz, Michael I. Jordan, and Pieter Abbeel. Trust region policy optimization. In *Proc. of the 32nd International Conference on Machine Learning (ICML)*, pages 1889–1897, 2015. 12

John Schulman, Filip Wolski, Prafulla Dhariwal, Alec Radford, and Oleg Klimov. Proximal policy optimization algorithms. *ArXiv Preprint ArXiv:1707.06347*, 2017. 12

Ming Tan. Multi-agent reinforcement learning: Independent vs. cooperative agents. In *Proc. of the 10th International Conference on Machine Learning (ICML)*, pages 330–337, 1993. DOI: 10.1016/b978-1-55860-307-3.50049-6 13, 45

Martin J. Osborne and Ariel Rubinstein. *A Course in Game Theory*. MIT Press, 1994. https://arielrubinstein.tau.ac.il/books/GT.pdf 13, 14, 15

David L. Poole and Alan K. Mackworth. *Artificial Intelligence: Foundations of Computational Agents*. Cambridge University Press, 2017. https://artint.info/2e/html/ArtInt2e.html DOI: 10.1017/9781108164085 14

Yoav Shoham and Kevin Leyton-Brown. *Multiagent Systems: Algorithmic, Game-Theoretic, and Logical Foundations*. Cambridge University Press, 2009. http://masfoundations.org DOI: 10.1017/cbo9780511811654 14, 15

Michael Bowling and Manuela Veloso. An analysis of stochastic game theory for multiagent reinforcement learning. *Technical Report*, Computer Science Department, Carnegie Mellon University, 2000. 15

Yujing Hu, Yang Gao, and Bo An. Learning in multi-agent systems with sparse interactions by knowledge transfer and game abstraction. In *Proc. of the 14th International Conference on Autonomous Agents and Multiagent Systems (AAMAS)*, pages 753–761, 2015a. 16, 34, 37, 66

Yujing Hu, Yang Gao, and Bo An. Accelerating multiagent reinforcement learning by equilibrium transfer. *IEEE Transactions on Cybernetics*, 45(7):1289–1302, 2015b. DOI: 10.1109/tcyb.2014.2349152 16

Liviu Panait and Sean Luke. Cooperative multi-agent learning: The state-of-the-art. *Autonomous Agents and Multiagent Systems*, 11(3):387–434, 2005. DOI: 10.1007/s10458-005-2631-2 16

Martin Lauer and Martin Riedmiller. An algorithm for distributed reinforcement learning in cooperative multi-agent systems. In *Proc. of the 17th International Conference on Machine Learning (ICML)*, pages 535–542, 2000. 17

Yujing Hu, Yang Gao, and Bo An. Multiagent reinforcement learning with unshared value functions. *IEEE Transactions on Cybernetics*, 45(4):647–662, 2015c. DOI: 10.1109/tcyb.2014.2332042 17

Junling Hu and Michael P. Wellman. Nash Q-learning for general-sum stochastic games. *Journal of Machine Learning Research (JMLR)*, 4:1039–1069, 2003. 17

Eric Sodomka, Elizabeth Hilliard, Michael L. Littman, and Amy Greenwald. Coco-Q: Learning in stochastic games with side payments. In *Proc. of the 30th International Conference on Machine Learning (ICML)*, 28:1471–1479, 2013. 17

Jakob N. Foerster, Richard Y. Chen, Maruan Al-Shedivat, Shimon Whiteson, Pieter Abbeel, and Igor Mordatch. Learning with opponent-learning awareness. In *Proc. of the 17th International Conference on Autonomous Agents and Multiagent Systems (AAMAS)*, pages 122–130, 2018a. 17

Michael L. Littman. Markov games as a framework for multi-agent reinforcement learning. In *Proc. of the 11th International Conference on Machine Learning (ICML)*, pages 157–163, 1994. DOI: 10.1016/b978-1-55860-335-6.50027-1 17, 67, 68

Ryan Lowe, Yi Wu, Aviv Tamar, Jean Harb, Pieter Abbeel, and Igor Mordatch. Multi-agent actor-critic for mixed cooperative-competitive environments. *Neural Information Processing Systems (NIPS)*, 2017. 17, 67

Pablo Hernandez-Leal, Michael Kaisers, Tim Baarslag, and Enrique Munoz de Cote. A survey of learning in multiagent environments: Dealing with non-stationarity. *ArXiv Preprint ArXiv:1707.09183*, 2017. 18

Jakob Foerster, Gregory Farquhar, Triantafyllos Afouras, Nantas Nardelli, and Shimon Whiteson. Counterfactual multi-agent policy gradients. In *Proc. of the 32nd AAAI Conference on Artificial Intelligence*, 2018b. 18

Xueguang Lyu, Yuchen Xiao, Brett Daley, and Christopher Amato. Contrasting centralized and decentralized critics in multi-agent reinforcement learning. In *Proc. of the 20th International Conference on Autonomous Agents and MultiAgent Systems (AAMAS)*, 2021. 18

Francisco S. Melo and Manuela Veloso. Decentralized MDPs with sparse interactions. *Artificial Intelligence*, 175(11):1757–1789, 2011. DOI: 10.1016/j.artint.2011.05.001 18

Y-M. De Hauwere, Peter Vrancx, and Ann Nowé. Learning multi-agent state space representations. In *Proc. of the 9th International Conference on Autonomous Agents and Multiagent Systems (AAMAS)*, pages 715–722, 2010. 18

Tom Croonenborghs, Karl Tuyls, Jan Ramon, and Maurice Bruynooghe. Multi-agent relational reinforcement learning. In *Learning and Adaption in Multi-Agent Systems*, pages 192–206, 2005. DOI: 10.1007/11691839_12 18, 38

Felipe Leno Da Silva, Ruben Glatt, and Anna H. R. Costa. MOO-MDP: An object-oriented representation for cooperative multiagent reinforcement learning. *IEEE Transactions on Cybernetics*, 49(2):567–579, 2019. DOI: 10.1109/tcyb.2017.2781130 18, 38, 51

Sinno J. Pan and Qiang Yang. A survey on transfer learning. *IEEE Transactions on Knowledge and Data Engineering*, 22(10):1345–1359, 2010. DOI: 10.1109/tkde.2009.191 19

Fernando Fernández and Manuela Veloso. Probabilistic policy reuse in a reinforcement learning agent. In *Proc. of the 5th International Joint Conference on Autonomous Agents and Multiagent Systems (AAMAS)*, pages 720–727, 2006. DOI: 10.1145/1160633.1160762 25

Sebastian Thrun and Tom M. Mitchell. Lifelong robot learning. *Robotics and Autonomous Systems*, 15(1–2):25–46, 1995. DOI: 10.1016/0921-8890(95)00004-y 25

Faraz Torabi, Garrett Warnell, and Peter Stone. Recent advances in imitation learning from observation. In *Proc. of the 28th International Joint Conference on Artificial Intelligence (IJCAI)*, pages 6325–6331, 2019. DOI: 10.24963/ijcai.2019/882 26

Sanmit Narvekar, Bei Peng, Matteo Leonetti, Jivko Sinapov, Matthew E. Taylor, and Peter Stone. Curriculum learning for reinforcement learning domains: A framework and survey. *Journal of Machine Learning Research (JMLR)*, 21(181):1–50, 2020. 26

Peter Stone, Gal A. Kaminka, Sarit Kraus, and Jeffrey S. Rosenschein. Ad hoc autonomous agent teams: Collaboration without pre-coordination. In *Proc. of the 24th AAAI Conference on Artificial Intelligence*, pages 1504–1509, 2010. 33, 35, 76

Stephen Kelly and Malcolm I. Heywood. Knowledge transfer from keepaway soccer to half-field offense through program symbiosis: Building simple programs for a complex task. In *Proc. of the 17th Conference on Genetic and Evolutionary Computation (GECCO)*, pages 1143–1150, 2015. DOI: 10.1145/2739480.2754798 33, 34, 36, 68

Bikramjit Banerjee and Peter Stone. General game learning using knowledge transfer. In *Proc. of the 20th International Joint Conference on Artificial Intelligence (IJCAI)*, pages 672–677, 2007. 34, 35

Samuel Barrett and Peter Stone. Cooperating with unknown teammates in complex domains: A robot soccer case study of ad hoc teamwork. In *Proc. of the 29th AAAI Conference on Artificial Intelligence*, pages 2010–2016, 2015. 34, 35

Pablo Hernandez-Leal, Yusen Zhan, Matthew E. Taylor, L. Enrique Sucar, and Enrique Munoz de Cote. Efficiently detecting switches against non-stationary opponents. *Autonomous Agents and Multi-Agent Systems*, 31(4):767–789, July 2017. DOI: 10.1007/s10458-016-9352-6 36

Pablo Hernandez-Leal and Michael Kaisers. Towards a fast detection of opponents in repeated stochastic games. In *Proc. of the 1st Workshop on Transfer in Reinforcement Learning (TiRL)*, 2017. DOI: 10.1007/978-3-319-71682-4_15 34, 36

Stefano V. Albrecht and Peter Stone. Autonomous agents modelling other agents: A comprehensive survey and open problems. *Artificial Intelligence*, 258:66–95, 2018. DOI: 10.1016/j.artint.2018.01.002 36

Yaqing Hou, Yew-Soon Ong, Jing Tang, and Yifeng Zeng. Evolutionary multiagent transfer learning with model-based opponent behavior prediction. *IEEE Transactions on Systems, Man, and Cybernetics: Systems*, pages 1–15, 2019. DOI: 10.1109/tsmc.2019.2958846 34, 36

Peter Vrancx, Yann-Michaël De Hauwere, and Ann Nowé. Transfer learning for multi-agent coordination. In *Proc. of the 3rd International Conference on Agents and Artificial Intelligence (ICAART)*, pages 263–272, 2011. 34, 37

L. Zhou, P. Yang, C. Chen, and Y. Gao. Multiagent reinforcement learning with sparse interactions by negotiation and knowledge transfer. *IEEE Transactions on Cybernetics*, 47(5):1238–1250, 2016. DOI: 10.1109/tcyb.2016.2543238 34, 37

Kristian Kersting, Martijn van Otterlo, and Luc De Raedt. Bellman goes relational. In *Proc. of the 21st International Conference on Machine Learning (ICML)*, pages 465–472, 2004. DOI: 10.1145/1015330.1015401 37

Carlos Diuk, Andre Cohen, and Michael L. Littman. An object-oriented representation for efficient reinforcement learning. In *Proc. of the 26th International Conference on Machine Learning (ICML)*, pages 240–247, 2008. DOI: 10.1145/1390156.1390187 37, 68

Marcelo Li Koga, Valdinei Freire da Silva, Fabio Gagliardi Cozman, and Anna Helena Reali Costa. Speeding-up reinforcement learning through abstraction and transfer learning. In *Proc. of the 12th International Conference on Autonomous Agents and Multiagent Systems (AAMAS)*, pages 119–126, 2013. 34, 38

Valdinei Freire and Anna Helena Reali Costa. Comparative analysis of abstract policies to transfer learning in robotics navigation. In *AAAI Workshop on Knowledge, Skill, and Behavior Transfer in Autonomous Robots*, pages 9–15, 2015. 34, 38

M. L. Koga, V. F. da Silva, and A. H. R. Costa. Stochastic abstract policies: Generalizing knowledge to improve reinforcement learning. *IEEE Transactions on Cybernetics*, 45(1):77–88, 2015. DOI: 10.1109/tcyb.2014.2319733 34, 38

Felipe Leno Da Silva and Anna Helena Reali Costa. Towards zero-shot autonomous inter-task mapping through object-oriented task description. In *Proc. of the 1st Workshop on Transfer in Reinforcement Learning (TiRL)*, 2017a. 34, 38, 66

Scott Proper and Prasad Tadepalli. Multiagent transfer learning via assignment-based decomposition. In *Proc. of the 8th International Conference on Machine Learning and Applications (ICMLA)*, pages 345–350, 2009. DOI: 10.1109/icmla.2009.59 34, 38

Alexander Braylan and Risto Miikkulainen. Object-model transfer in the general video game domain. In *Proc. of the 12th AAAI Conference on Artificial Intelligence and Interactive Digital Entertainment (AIIDE)*, pages 136–142, 2016. 34, 39

Tesca Fitzgerald, Kalesha Bullard, Andrea Thomaz, and Ashok Goel. Situated mapping for transfer learning. In *Proc. of the 4th Annual Conference on Advances in Cognitive Systems*, pages 1–14, 2016. 39

Reinaldo A. C. Bianchi, Luiz A. Celiberto-Jr., Paulo E. Santos, Jackson P. Matsuura, and Ramon Lopez de Mantaras. Transferring knowledge as heuristics in reinforcement learning: A case-based approach. *Artificial Intelligence*, 226:102–121, 2015. DOI: 10.1016/j.artint.2015.05.008 39

Reinaldo Bianchi, Raquel Ros, and Ramon Lopez de Mantaras. Improving reinforcement learning by using case based heuristics. *Case-Based Reasoning Research and Development*, pages 75–89, 2009. DOI: 10.1007/978-3-642-02998-1_7 34, 39, 67

Georgios Boutsioukis, Ioannis Partalas, and Ioannis Vlahavas. Transfer learning in multi-agent reinforcement learning domains. In *Proc. of the 9th European Workshop on Reinforcement Learning*, 2011. DOI: 10.1007/978-3-642-29946-9_25 34, 40

Sabre Didi and Geoff Nitschke. Multi-agent behavior-based policy transfer. In Giovanni Squillero and Paolo Burelli, Eds., *Proc. of the 19th European Conference on Applications of Evolutionary Computation (EvoApplications)*, pages 181–197, 2016. DOI: 10.1007/978-3-319-31153-1_13 34, 40

Kenneth O. Stanley and Risto Miikkulainen. Evolving neural networks through augmenting topologies. *Evolutionary Computation*, 10(2):99–127, 2002. DOI: 10.1162/106365602320169811 40

Sanmit Narvekar, Jivko Sinapov, Matteo Leonetti, and Peter Stone. Source task creation for curriculum learning. In *Proc. of the 15th International Conference on Autonomous Agents and Multiagent Systems (AAMAS)*, pages 566–574, 2016. 34, 40

Yoshua Bengio, Jérôme Louradour, Ronan Collobert, and Jason Weston. Curriculum learning. In *Proc. of the 26th International Conference on Machine Learning (ICML)*, pages 41–48, 2009. DOI: 10.1145/1553374.1553380 40

Maxwell Svetlik, Matteo Leonetti, Jivko Sinapov, Rishi Shah, Nick Walker, and Peter Stone. Automatic curriculum graph generation for reinforcement learning agents. In *Proc. of the 31st AAAI Conference on Artificial Intelligence*, pages 2590–2596, 2017. 34, 40, 69

Felipe Leno Da Silva and Anna Helena Reali Costa. Object-oriented curriculum generation for reinforcement learning. In *Proc. of the 17th International Conference on Autonomous Agents and Multiagent Systems (AAMAS)*, pages 1026–1034, 2018. 34, 41

Sanmit Narvekar, Jivko Sinapov, and Peter Stone. Autonomous task sequencing for customized curriculum design in reinforcement learning. In *Proc. of the 26th International Joint Conference on Artificial Intelligence (IJCAI)*, pages 2536–2542, 2017. DOI: 10.24963/ijcai.2017/353 34, 41, 66

Carlos Florensa, David Held, Markus Wulfmeier, Michael Zhang, and Pieter Abbeel. Reverse curriculum generation for reinforcement learning. In Sergey Levine, Vincent Vanhoucke, and Ken Goldberg, Eds., *Proc. of the 1st Conference on Robot Learning (CoRL)*, vol. 78 of *Proc. of Machine Learning Research*, pages 482–495, 2017. 34, 41

Michael G. Madden and Tom Howley. Transfer of experience between reinforcement learning environments with progressive difficulty. *Artificial Intelligence Review*, 21(3):375–398, 2004. DOI: 10.1023/b:aire.0000036264.95672.64 34, 41, 54, 66, 78

Lerrel Pinto, James Davidson, Rahul Sukthankar, and Abhinav Gupta. Robust adversarial reinforcement learning. In *Proc. of the 34th International Conference on Machine Learning (ICML)*, pages 2817–2826, 2017. 34, 41, 70

Akshat Agarwal, Sumit Kumar, Katia Sycara, and Michael Lewis. Learning transferable cooperative behavior in multi-agent teams. In *Proc. of the 19th International Conference on Autonomous Agents and MultiAgent Systems (AAMAS)*, pages 1741–1743, 2020. 35, 42

Heechang Ryu, Hayong Shin, and Jinkyoo Park. Multi-agent actor-critic with hierarchical graph attention network. In *Proc. of the 34th AAAI Conference on Artificial Intelligence*, 2020. DOI: 10.1609/aaai.v34i05.6214 35, 42

François-Xavier Devailly, Denis Larocque, and Laurent Charlin. IG-RL: Inductive graph reinforcement learning for massive-scale traffic signal control. *ArXiv Preprint ArXiv:2003.05738*, 2020. DOI: 10.1109/tits.2021.3070835 35, 42

Alexander A. Sherstov and Peter Stone. Improving action selection in MDP's via knowledge transfer. In *Proc. of the 20th AAAI Conference on Artificial Intelligence*, pages 1024–1029, 2005. 35, 43

George Konidaris and Andrew Barto. Autonomous shaping: Knowledge transfer in reinforcement learning. In *Proc. of the 23rd International Conference on Machine Learning (ICML)*, pages 489–496, 2006. DOI: 10.1145/1143844.1143906 35, 43

E. Chalmers, E. B. Contreras, B. Robertson, A. Luczak, and A. Gruber. Learning to predict consequences as a method of knowledge transfer in reinforcement learning.

IEEE Transactions on Neural Networks and Learning Systems, 26(6):2259–2270, 2017. DOI: 10.1109/tnnls.2017.2690910 35, 43

Enrique Munoz de Cote, Esteban O. Garcia, and Eduardo F. Morales. Transfer learning by prototype generation in continuous spaces. *Adaptive Behavior*, 24(6):464–478, 2016. DOI: 10.1177/1059712316664570 35, 43

Shane Griffith, Kaushik Subramanian, Jonathan Scholz, Charles L. Isbell, and Andrea L. Thomaz. Policy shaping: Integrating human feedback with reinforcement learning. In *Advances in Neural Information Processing Systems (NIPS)*, pages 2625–2633, 2013. 45, 46

Thomas Cederborg, Ishaan Grover, Charles L. Isbell, and Andrea Lockerd Thomaz. Policy shaping with human teachers. In *Proc. of the 24th International Joint Conference on Artificial Intelligence (IJCAI)*, pages 3366–3372, 2015. 48

Lisa Torrey and Matthew E. Taylor. Teaching on a budget: Agents advising agents in reinforcement learning. In *Proc. of 12th the International Conference on Autonomous Agents and MultiAgent Systems (AAMAS)*, pages 1053–1060, 2013. 46, 48

Matthew E. Taylor, Nicholas Carboni, Anestis Fachantidis, Ioannis P. Vlahavas, and Lisa Torrey. Reinforcement learning agents providing advice in complex video games. *Connection Science*, 26(1):45–63, 2014a. DOI: 10.1080/09540091.2014.885279 48, 68

Yusen Zhan, Haitham Bou-Ammar, and Matthew E. Taylor. Theoretically-grounded policy advice from multiple teachers in reinforcement learning settings with applications to negative transfer. In *Proc. of the 25th International Joint Conference on Artificial Intelligence (IJCAI)*, pages 2315–2321, 2016. 46, 48

Ofra Amir, Ece Kamar, Andrey Kolobov, and Barbara Grosz. Interactive teaching strategies for agent training. In *Proc. of the 25th International Joint Conference on Artificial Intelligence (IJCAI)*, pages 804–811, 2016. 46, 48

Ercüment Ilhan, Jeremy Gow, and Diego Perez-Liebana. Teaching on a budget in multi-agent deep reinforcement learning. In *IEEE Conference on Games (CoG)*, pages 1–8, 2019. DOI: 10.1109/cig.2019.8847988 46, 48

Felipe Leno Da Silva, Pablo Hernandez-Leal, Bilal Kartal, and Matthew E. Taylor. Uncertainty-aware action advising for deep reinforcement learning agents. In *Proc. of the 34th AAAI Conference on Artificial Intelligence*, 2020b. DOI: 10.1609/aaai.v34i04.6036 46, 48

Changxi Zhu, Yi Cai, Ho-fung Leung, and Shuyue Hu. Learning by reusing previous advice in teacher-student paradigm. In *Proc. of the 19th International Conference on Autonomous Agents and MultiAgent Systems (AAMAS)*, pages 1674–1682, 2020. 46, 49

Matthieu Zimmer, Paolo Viappiani, and Paul Weng. Teacher-student framework: A reinforcement learning approach. In *Workshop on Autonomous Robots and Multirobot Systems at AAMAS*, 2014. 49

Anestis Fachantidis, Matthew E. Taylor, and Ioannis Vlahavas. Learning to teach reinforcement learning agents. *Machine Learning and Knowledge Extraction*, 1(1):2018. DOI: 10.3390/make1010002 46, 49

Shayegan Omidshafiei, Dong-Ki Kim, Miao Liu, Gerald Tesauro, Matthew Riemer, Christopher Amato, Murray Campbell, and Jonathan P. How. Learning to teach in cooperative multiagent reinforcement learning. In *Workshop on Lifelong Learning: A. Reinforcement Learning Approach*, 2018. DOI: 10.1609/aaai.v33i01.33016128 46, 49

Dong-Ki Kim, Miao Liu, Shayegan Omidshafiei, Sebastian Lopez-Cot, Matthew Riemer, Golnaz Habibi, Gerald Tesauro, Sami Mourad, Murray Campbell, and Jonathan P. How. Learning hierarchical teaching policies for cooperative agents. In *Proc. of the 19th International Conference on Autonomous Agents and MultiAgent Systems (AAMAS)*, pages 620–628, 2020. 46, 49

Richard Maclin, Jude W. Shavlik, and Pack Kaelbling. Creating advice-taking reinforcement learners. In *Machine Learning*, 22:251–281, 1996. DOI: 10.1007/bf00114730 46, 50

W. Bradley Knox and Peter Stone. Interactively shaping agents via human reinforcement: The TAMER framework. In *Proc. of the 5th International Conference on Knowledge Capture*, pages 9–16, September 2009. DOI: 10.1145/1597735.1597738 46, 50

Kshitij Judah, Saikat Roy, Alan Fern, and Thomas G. Dietterich. Reinforcement learning via practice and critique advice. In *Proc. of the 24th AAAI Conference on Artificial Intelligence*, pages 481–486, 2010. 46, 50

Bei Peng, James MacGlashan, Robert Loftin, Michael L. Littman, David L. Roberts, and Matthew E. Taylor. A need for speed: Adapting agent action speed to improve task learning from non-expert humans. In *Proc. of the 15th International Conference on Autonomous Agents and Multiagent Systems (AAMAS)*, pages 957–965, 2016a. 46, 51, 66, 79

James MacGlashan, Mark K. Ho, Robert Loftin, Bei Peng, Guan Wang, David L. Roberts, Matthew E. Taylor, and Michael L. Littman. Interactive learning from policy-dependent human feedback. In *Proc. of the 34th International Conference on Machine Learning (ICML)*, pages 2285–2294, 2017. 46, 51

David Abel, John Salvatier, Andreas Stuhlmüller, and Owain Evans. Agent-agnostic human-in-the-loop reinforcement learning. In *Proc. of the NIPS Future of Interactive Learning Machines Workshop*, 2016. 46, 51

Ariel Rosenfeld, Matthew E. Taylor, and Sarit Kraus. Leveraging human knowledge in tabular reinforcement learning: A study of human subjects. In *Proc. of the 26th International Joint Conference on Artificial Intelligence (IJCAI)*, pages 3823–3830, 2017. DOI: 10.24963/ijcai.2017/534 46, 51

Samantha Krening, Brent Harrison, Karen M. Feigh, Charles L. Isbell, Mark Riedl, and Andrea Thomaz. Learning from explanations using sentiment and advice in RL. *IEEE Transactions on Cognitive and Developmental Systems*, 9(1):44–55, 2017. DOI: 10.1109/tcds.2016.2628365 46, 51, 69

Luis C. Cobo, Charles L. Isbell, and Andrea L. Thomaz. Object focused Q-learning for autonomous agents. In *Proc. of 12th the International Conference on Autonomous Agents and MultiAgent Systems (AAMAS)*, pages 1061–1068, 2013. 51

Travis Mandel, Yun-En Liu, Emma Brunskill, and Zoran Popovic. Where to add actions in human-in-the-loop reinforcement learning. In *Proc. of the 31st AAAI Conference on Artificial Intelligence*, pages 2322–2328, 2017. 46, 52

Stefan Schaal. Learning from demonstration. In *Advances in Neural Information Processing Systems (NIPS)*, pages 1040–1046, 1997. DOI: 10.1016/j.robot.2004.03.003 46, 52, 70

J. Zico Kolter, Pieter Abbeel, and Andrew Y. Ng. Hierarchical apprenticeship learning with application to quadruped locomotion. In *Advances in Neural Information Processing Systems (NIPS)*, pages 769–776, 2008. 46, 52, 66, 70

Sonia Chernova and Manuela Veloso. Confidence-based policy learning from demonstration using Gaussian mixture models. In *Proc. of the 6th International Joint Conference on Autonomous Agents and Multiagent Systems (AAMAS)*, page 233:1–233:8, 2007. DOI: 10.1145/1329125.1329407 53

Sonia Chernova and Manuela Veloso. Multi-thresholded approach to demonstration selection for interactive robot learning. In *Proc. of the 3rd ACM/IEEE International Conference on Human-Robot Interaction (HRI)*, pages 225–232, 2008. DOI: 10.1145/1349822.1349852 53

Sonia Chernova and Manuela Veloso. Interactive policy learning through confidence-based autonomy. *Journal of Artificial Intelligence Research (JAIR)*, 34(1):1–25, 2009. DOI: 10.1613/jair.2584 46, 53

Kshitij Judah, Alan Fern, and Thomas G. Dietterich. Active imitation learning via reduction to I.I.D. active learning. In *Proc. of the 28th Conference on Uncertainty in Artificial Intelligence (UAI)*, page 428,437, 2012. 53

Kshitij Judah, Alan P. Fern, Thomas G. Dietterich, and Prasad Tadepalli. Active imitation learning: Formal and practical reductions to I.I.D. learning. *Journal of Machine Learning Research (JMLR)*, 15(1):3925–3963, 2014. 46, 53

Roberto Capobianco. Robust and incremental robot learning by imitation. In *Proc. of the Doctoral Workshop in Artificial Intelligence (DWAI)*, 2014. 53

Matthew E. Taylor, Halit Bener Suay, and Sonia Chernova. Integrating reinforcement learning with human demonstrations of varying ability. In *The 10th International Conference on Autonomous Agents and Multiagent Systems (AAMAS)*, 2011. 53

Thomas J. Walsh, Daniel K. Hewlett, and Clayton T. Morrison. Blending autonomous exploration and apprenticeship learning. In *Advances in Neural Information Processing Systems (NIPS)*, pages 2258–2266. 2011. 46, 53

Lihong Li, Michael L. Littman, Thomas J. Walsh, and Alexander L. Strehl. Knows what it knows: A framework for self-aware learning. *Machine Learning*, 82(3):399–443, 2011. DOI: 10.1007/s10994-010-5225-4 54

Tim Brys, Anna Harutyunyan, Halit Bener Suay, Sonia Chernova, Matthew E. Taylor, and Ann Nowé. Reinforcement learning from demonstration through shaping. In *Proc. of the 24th International Joint Conference on Artificial Intelligence (IJCAI)*, pages 3352–3358, 2015a. 46, 54

Guo-fang Wang, Zhou Fang, Ping Li, and Bo Li. Transferring knowledge from human-demonstration trajectories to reinforcement learning. *Transactions of the Institute of Measurement and Control*, 40(1):94–101, 2016. DOI: 10.1177/0142331216649655 46, 54, 78

Kaushik Subramanian, Charles L. Isbell Jr., and Andrea L. Thomaz. Exploration from demonstration for interactive reinforcement learning. In *Proc. of the 15th International Conference on Autonomous Agents and Multiagent Systems (AAMAS)*, pages 447–456, 2016. 46, 54, 69

Zhaodong Wang and Matthew E. Taylor. Improving reinforcement learning with confidence-based demonstrations. In *Proc. of the 26th International Joint Conference on Artificial Intelligence (IJCAI)*, pages 3027–3033, 2017. DOI: 10.24963/ijcai.2017/422 46, 54

Bikramjit Banerjee, Syamala Vittanala, and Matthew E. Taylor. Team learning from human demonstration with coordination confidence. *The Knowledge Engineering Review*, 34, 2019. DOI: 10.1017/s0269888919000043 46, 54

Marco Tamassia, Fabio Zambetta, William Raffe, Florian Mueller, and Xiaodong Li. Learning options from demonstrations: A pac-man case study. *IEEE Transactions on Computational Intelligence and AI in Games*, 10(1):91–96, 2017. DOI: 10.1109/tciaig.2017.2658659 46, 55

Tianpei Yang, Weixun Wang, Hongyao Tang, Jianye Hao, Zhaopeng Meng, Wulong Liu, Yujing Hu, and Yingfeng Chen. Learning when to transfer among agents: An efficient multiagent transfer learning framework. *ArXiv Preprint ArXiv:2002.08030*, 2020. 46, 55

Bob Price and Craig Boutilier. Implicit imitation in multiagent reinforcement learning. In *Proc. of the 16th International Conference on Machine Learning (ICML)*, pages 325–334, 1999. 55

Bob Price and Craig Boutilier. Accelerating reinforcement learning through implicit imitation. *Journal of Artificial Intelligence Research (JAIR)*, 19:569–629, 2003. DOI: 10.1613/jair.898 47, 55

Aaron P. Shon, Deepak Verma, and Rajesh P. N. Rao. Active imitation learning. In *Proc. of the 21st AAAI Conference on Artificial Intelligence*, pages 756–762, 2007. 47, 56

Hoang Minh Le, Yisong Yue, and Peter Carr. Coordinated multi-agent imitation learning. In *Proc. of the 34th International Conference on Machine Learning (ICML)*, pages 1995–2003, 2017. 47, 56

Tatsuya Sakato, Motoyuki Ozeki, and Natsuki Oka. Learning through imitation and reinforcement learning: Toward the acquisition of painting motions. In *Proc. of the 3rd International Conference on Advanced Applied Informatics (IIAI)*, pages 873–880, 2014. DOI: 10.1109/iiai-aai.2014.174 47, 56, 70

Faraz Torabi, Garrett Warnell, and Peter Stone. Behavioral cloning from observation. In *Proc. of the 27th International Joint Conference on Artificial Intelligence (IJCAI)*, 2018. DOI: 10.24963/ijcai.2018/687 47, 56

Andrew Y. Ng, Daishi Harada, and Stuart Russell. Policy invariance under reward transformations: Theory and application to reward shaping. In *Proc. of the 16th International Conference on Machine Learning (ICML)*, pages 278–287, 1999. 57

Sam Devlin and Daniel Kudenko. Theoretical considerations of potential-based reward shaping for multi-agent systems. In *The 10th International Conference on Autonomous Agents and Multiagent Systems (AAMAS)*, pages 225–232, 2011. 57

Eric Wiewiora, Garrison W. Cottrell, and Charles Elkan. Principled methods for advising reinforcement learning agents. In *Proc. of the 20th International Conference on Machine Learning (ICML)*, pages 792–799, 2003. 47, 57

Sam Devlin, Logan Yliniemi, Daniel Kudenko, and Kagan Tumer. Potential-based difference rewards for multiagent reinforcement learning. In *Proc. of the 13th International Conference on Autonomous Agents and Multiagent Systems (AAMAS)*, pages 165–172, 2014. 47, 57, 80

Tim Brys, Anna Harutyunyan, Matthew E. Taylor, and Ann Nowé. Policy transfer using reward shaping. In *Proc. of the 14th International Conference on Autonomous Agents and Multiagent Systems (AAMAS)*, pages 181–188, 2015b. 57

Halit Bener Suay, Tim Brys, Matthew E. Taylor, and Sonia Chernova. Learning from demonstration for shaping through inverse reinforcement learning. In *Proc. of the 15th International Conference on Autonomous Agents and Multiagent Systems (AAMAS)*, pages 429–437, 2016. 47, 57

Abhishek Gupta, Coline Devin, Yuxuan Liu, Pieter Abbeel, and Sergey Levine. Learning invariant feature spaces to transfer skills with reinforcement learning. In *Proc. of the 5th International Conference on Learning Representations (ICLR)*, 2017a. 47, 57, 70

Paniz Behboudian, Yash Satsangic, Matthew E. Taylor, Anna Harutyunyan, and Michael Bowling. Useful policy invariant shaping from arbitrary advice. In *AAMAS Adaptive Learning Agents (ALA) Workshop*, 2020. 47, 57

Danilo H. Perico and Reinaldo A. C. Bianchi. Use of heuristics from demonstrations to speed up reinforcement learning. In *Proc. of the 12th Brazilian Symposium on Intelligent Automation (SBAI)*, 2013. 47, 57

Reinaldo A. C. Bianchi, Murilo F. Martins, Carlos H. C. Ribeiro, and Anna H. R. Costa. Heuristically-accelerated multiagent reinforcement learning. *IEEE Transactions on Cybernetics*, 44(2):252–265, 2014. DOI: 10.1109/tcyb.2013.2253094 47, 58

Deepak Ramachandran and Eyal Amir. Bayesian inverse reinforcement learning. In *Proc. of the 20th International Joint Conference on Artificial Intelligence (IJCAI)*, pages 2586–2591, 2007. 58

Manuel Lopes, Francisco Melo, and Luis Montesano. Active learning for reward estimation in inverse reinforcement learning. In *Joint European Conference on Machine Learning and Knowledge Discovery in Databases (ECML/PKDD)*, pages 31–46, 2009. DOI: 10.1007/978-3-642-04174-7_3 47, 58

Yuchen Cui and Scott Niekum. Active reward learning from critiques. In *IEEE International Conference on Robotics and Automation (ICRA)*, pages 6907–6914, 2018. DOI: 10.1109/icra.2018.8460854 47, 58

Tummalapalli Sudhamsh Reddy, Vamsikrishna Gopikrishna, Gergely Zaruba, and Manfred Huber. Inverse reinforcement learning for decentralized non-cooperative multiagent systems. In *Proc. of the IEEE International Conference on Systems, Man, and Cybernetics (SMC)*, pages 1930–1935, 2012. DOI: 10.1109/icsmc.2012.6378020 47, 58, 59

Sriraam Natarajan, Gautam Kunapuli, Kshitij Judah, Prasad Tadepalli, Kristian Kersting, and Jude Shavlik. Multi-agent inverse reinforcement learning. In *Proc. of the 9th International Conference on Machine Learning and Applications (ICMLA)*, pages 395–400, IEEE, 2010. DOI: 10.1109/icmla.2010.65 47, 59, 82

Xiaomin Lin, Peter A. Beling, and Randy Cogill. Multiagent inverse reinforcement learning for two-person zero-sum games. *IEEE Transactions on Games*, 10(1):56–68, 2018. DOI: 10.1109/tciaig.2017.2679115 47, 59, 82

Kyriacos Shiarlis, Joao Messias, and Shimon Whiteson. Inverse reinforcement learning from failure. In *Proc. of the 15th International Conference on Autonomous Agents and Multiagent Systems (AAMAS)*, pages 1060–1068, 2016. 47, 59

Voot Tangkaratt, Bo Han, Mohammad Emtiyaz Khan, and Masashi Sugiyama. Variational imitation learning with diverse-quality demonstrations. In *Proc. of the 37th International Conference on Machine Learning (ICML)*, 2020a. 47, 59

Bei Peng, James MacGlashan, Robert Loftin, Michael L. Littman, David L. Roberts, and Matthew E. Taylor. An empirical study of non-expert curriculum design for machine learners. In *Proc. of the IJCAI Interactive Machine Learning Workshop*, 2016b. 47, 59

Tambet Matiisen, Avital Oliver, Taco Cohen, and John Schulman. Teacher-student curriculum learning. In *Deep Reinforcement Learning Symposium at NIPS*, 2017. DOI: 10.1109/tnnls.2019.2934906 47, 60

Sainbayar Sukhbaatar, Ilya Kostrikov, Arthur Szlam, and Rob Fergus. Intrinsic motivation and automatic curricula via asymmetric self-play. In *Proc. of the 6th International Conference on Learning Representations (ICLR)*, 2018. 47, 60

Alvaro Ovalle Castaneda. Deep reinforcement learning variants of multi-agent learning algorithms. Ph.D. thesis, University of Edinburgh, 2016. 60, 77

Jayesh K. Gupta, Maxim Egorov, and Mykel Kochenderfer. Cooperative multi-agent control using deep reinforcement learning. In *AAMAS Adaptive Learning Agents (ALA) Workshop*, 2017b. DOI: 10.1007/978-3-319-71682-4_5 60, 77

Ruben Glatt, Felipe Leno Da Silva, and Anna Helena Reali Costa. Towards knowledge transfer in deep reinforcement learning. In *Brazilian Conference on Intelligent Systems (BRACIS)*, pages 91–96, 2016. DOI: 10.1109/bracis.2016.027 60, 69, 77

Yunshu Du, Gabriel V. de la Cruz, James Irwin, and Matthew E. Taylor. Initial progress in transfer for deep reinforcement learning algorithms. In *Proc. of Deep Reinforcement Learning: Frontiers and Challenges Workshop at IJCAI*, 2016. 60, 69, 77

Jakob N. Foerster, Yannis M. Assael, Nando de Freitas, and Shimon Whiteson. Learning to communicate with deep multi-agent reinforcement learning. In *Conference on Neural Information Processing Systems (NIPS)*, 2016. 47, 60, 76, 77

Sainbayar Sukhbaatar, Arthur Szlam, and Rob Fergus. Learning multiagent communication with backpropagation. In *Conference on Neural Information Processing Systems (NIPS)*, 2016. 47, 60, 77

Coline Devin, Abhishek Gupta, Trevor Darrell, Pieter Abbeel, and Sergey Levine. Learning modular neural network policies for multi-task and multi-robot transfer. In *IEEE International Conference on Robotics and Automation (ICRA)*, pages 2169–2176, 2017. DOI: 10.1109/icra.2017.7989250 47, 61, 70

Gabriel V. de la Cruz, Yunshu Du, and Matthew E. Taylor. Pre-training with non-expert human demonstration for deep reinforcement learning. *The Knowledge Engineering Review*, 34:e10, 2019. DOI: 10.1017/s0269888919000055 47, 61

Shayegan Omidshafiei, Jason Pazis, Christopher Amato, Jonathan P. How, and John Vian. Deep decentralized multi-task multi-agent reinforcement learning under partial observability. In *Proc. of the 34th International Conference on Machine Learning (ICML)*, pages 2681–2690, 2017. 47, 61

Lucas Oliveira Souza, Gabriel de Oliveira Ramos, and Celia Ghedini Ralha. Experience sharing between cooperative reinforcement learning agents. In *Proc. of the 31st IEEE International Conference on Tools with Artificial Intelligence (ICTAI)*, pages 963–970, 2019. DOI: 10.1109/ictai.2019.00136 47, 61

Kwei-Herng Lai, Daochen Zha, Yuening Li, and Xia Hu. Dual policy distillation. In *Proc. of the 29th International Joint Conference on Artificial Intelligence (IJCAI)*, 2020. DOI: 10.24963/ijcai.2020/435 47, 61

Adam Taylor, Ivana Dusparic, Edgar Galvan-Lopez, Siobhan Clarke, and Vinny Cahill. Accelerating learning in multi-objective systems through transfer learning. In *International Joint Conference on Neural Networks (IJCNN)*, pages 2298–2305, 2014b. DOI: 10.1109/ijcnn.2014.6889438 47, 62, 70

Ivana Dusparic and Vinny Cahill. Distributed W-learning: Multi-policy optimization in self-organizing systems. In *3rd IEEE International Conference on Self-Adaptive and Self-Organizing Systems (SASO)*, pages 20–29, 2009. DOI: 10.1109/saso.2009.23 62

Adam Taylor, Ivana Dusparic, Maxime Guériau, and Siobhán Clarke. Parallel transfer learning in multi-agent systems: What, when and how to transfer? In *International Joint Conference on Neural Networks (IJCNN)*, pages 1–8, 2019. DOI: 10.1109/ijcnn.2019.8851784 47, 62

Hitoshi Kono, Akiya Kamimura, Kohji Tomita, Yuta Murata, and Tsuyoshi Suzuki. Transfer learning method using ontology for heterogeneous multi-agent reinforcement learning. *International Journal of Advanced Computer Science and Applications (IJACSA)*, 5(10):156–164, 2014. DOI: 10.14569/ijacsa.2014.051022 47, 62, 80

Yanhai Xiong, Haipeng Chen, Mengchen Zhao, and Bo An. HogRider: Champion agent of microsoft Malmo collaborative AI challenge. In *Proc. of the 32nd AAAI Conference on Artificial Intelligence*, pages 4767–4774, 2018. 47, 62

Carlos Diuk. An object-oriented representation for efficient reinforcement learning. Ph.D. thesis, Rutgers University, 2009. DOI: 10.1145/1390156.1390187 66

Thomas G. Dietterich. Hierarchical reinforcement learning with the MAXQ value function decomposition. *Journal of Artificial Intelligence Research (JAIR)*, 13:227–303, 2000. DOI: 10.1613/jair.639 66

Hiroaki Kitano, Minoru Asada, Yasuo Kuniyoshi, Itsuki Noda, Eiichi Osawa, and Hitoshi Mat-subara. RoboCup: A challenge problem for AI. *AI Magazine*, 18(1):73–85, 1997. DOI: 10.1007/3-540-64473-3_46 67

RoboCup. RoboCup 2D simulation league. 2019. http://www.robocup.org/leagues/23 67

Michael W. Floyd, Babak Esfandiari, and Kevin Lam. A case-based reasoning approach to imitating RoboCup players. In *Proc. of the 21st International Florida Artificial Intelligence Research Society Conference (FLAIRS)*, pages 251–256, 2008. 67

Peter Stone, Richard S. Sutton, and Gregory Kuhlmann. Reinforcement learning for RoboCup-soccer keepaway. *Adaptive Behavior*, 13(3):165–188, 2005. DOI: 10.1177/105971230501300301 67, 68

Matthew Hausknecht, Prannoy Mupparaju, Sandeep Subramanian, Shivaram Kalyanakrishnan, and Peter Stone. Half field offense: An environment for multiagent learning and ad hoc teamwork. In *AAMAS Adaptive Learning Agents (ALA) Workshop*, May 2016. 67, 68

Christopher Berner, Greg Brockman, Brooke Chan, Vicki Cheung, Przemysław Dębiak, Christy Dennison, David Farhi, Quirin Fischer, Shariq Hashme, Chris Hesse, Rafal Józe-fowicz, Scott Gray, Catherine Olsson, Jakub Pachocki, Michael Petrov, Henrique Pondé de Oliveira Pinto, Jonathan Raiman, Tim Salimans, Jeremy Schlatter, Jonas Schneider, Szymon Sidor, Ilya Sutskever, Jie Tang, Filip Wolski, and Susan Zhang. Dota 2 with large scale deep reinforcement learning. *ArXiv Preprint ArXiv:1912.06680*, 2019. 69, 86

Josiah Hanna and Peter Stone. Grounded action transformation for robot learning in simulation. In *Proc. of the 31st AAAI Conference on Artificial Intelligence*, pages 3834–3840, 2017. 70

Emanuel Todorov, Tom Erez, and Yuval Tassa. MuJoCo: A physics engine for model-based control. In *IEEE/RSJ International Conference on Intelligent Robots and Systems*, pages 5026–5033, 2012. DOI: 10.1109/iros.2012.6386109 70

Claudine Badue, Rânik Guidolini, Raphael Vivacqua Carneiro, Pedro Azevedo, Vinicius B. Cardoso, Avelino Forechi, Luan Jesus, Rodrigo Berriel, Thiago M. Paix ao, Filipe Mutz, Lucas de Paula Veronese, Thiago Oliveira-Santos, and Alberto F. De Souza. Self-driving cars: A survey. *Expert Systems with Applications*, 165:113816, 2021. DOI: 10.1016/j.eswa.2020.113816 72

Ming Zhou, Jun Luo, Julian Villella, Yaodong Yang, David Rusu, Jiayu Miao, Weinan Zhang, Montgomery Alban, Iman Fadakar, Zheng Chen, Aurora Chongxi Huang, Ying Wen, Kimia Hassanzadeh, Daniel Graves, Dong Chen, Zhengbang Zhu, Nhat Nguyen, Mohamed Elsayed, Kun Shao, Sanjeevan Ahilan, Baokuan Zhang, Jiannan Wu, Zhengang Fu, Kasra Rezaee, Peyman Yadmellat, Mohsen Rohani, Nicolas Perez Nieves, Yihan Ni, Seyedershad Banijamali, Alexander Cowen Rivers, Zheng Tian, Daniel Palenicek, Haitham bou Ammar, Hongbo Zhang, Wulong Liu, Jianye Hao, and Jun Wang. SMARTS: Scalable multi-agent reinforcement learning training school for autonomous driving. *ArXiv Preprint ArXiv:2010.09776*, 2020. 72

Joel Z. Leibo, Edward Hughes, Marc Lanctot, and Thore Graepel. Autocurricula and the emergence of innovation from social interaction: A manifesto for multi-agent intelligence research. *ArXiv Preprint ArXiv:1903.00742*, 2019. 73

Scott M. Jordan, Yash Chandak, Daniel Cohen, Mengxue Zhang, and Philip S. Thomas. Evaluating the performance of reinforcement learning algorithms. In *Proc. of the 37th International Conference on Machine Learning (ICML)*, 2020. 75

Peter Vamplew, Richard Dazeley, Adam Berry, Rustam Issabekov, and Evan Dekker. Empirical evaluation methods for multiobjective reinforcement learning algorithms. *Machine Learning*, 84(1):51–80, 2011. DOI: 10.1007/s10994-010-5232-5 75

Felipe Leno Da Silva and Anna Helena Reali Costa. Accelerating multiagent reinforcement learning through transfer learning. In *Proc. of the 31st AAAI Conference on Artificial Intelligence*, pages 5034–5035, 2017b. 78

Ramya Ramakrishnan, Karthik Narasimhan, and Julie Shah. Interpretable transfer for reinforcement learning based on object similarities. In *Proc. of the IJCAI Interactive Machine Learning Workshop*, 2016. 79

Yaodong Yang, Rui Luo, Minne Li, Ming Zhou, Weinan Zhang, and Jun Wang. Mean field multi-agent reinforcement learning. In *Proc. of the 35th International Conference on Machine Learning (ICML)*, vol. 80 of *Proc. of Machine Learning Research*, pages 5571–5580, 2018. 81

Sriram Ganapathi Subramanian, Pascal Poupart, Matthew E. Taylor, and Nidhi Hegde. Multi type mean field reinforcement learning. In *Proc. of the 19th International Conference on Autonomous Agents and MultiAgent Systems (AAMAS)*, 2020. 81

Estefania Argente, Vicente Julian, and Vicente Botti. Multi-agent system development based on organizations. *Electronic Notes in Theoretical Computer Science*, 150(3):55–71, 2006. DOI: 10.1016/j.entcs.2006.03.005 82

Voot Tangkaratt, Bo Han, Mohammad Emtiyaz Khan, and Masashi Sugiyama. Policy teaching via environment poisoning: Training-time adversarial attacks against reinforcement learning. In *Proc. of the 37th International Conference on Machine Learning (ICML)*, 2020b. 82

James MacGlashan. Brown-UMBC reinforcement learning and planning (BURLAP), 2015. http://burlap.cs.brown.edu/index.html 86

Martín Abadi, Ashish Agarwal, Paul Barham, Eugene Brevdo, Zhifeng Chen, Craig Citro, Greg S. Corrado, Andy Davis, Jeffrey Dean, Matthieu Devin, Sanjay Ghemawat, Ian Goodfellow, Andrew Harp, Geoffrey Irving, Michael Isard, Yangqing Jia, Rafal Jozefowicz, Lukasz Kaiser, Manjunath Kudlur, Josh Levenberg, Dan Mané, Rajat Monga, Sherry Moore, Derek Murray, Chris Olah, Mike Schuster, Jonathon Shlens, Benoit Steiner, Ilya Sutskever, Kunal Talwar, Paul Tucker, Vincent Vanhoucke, Vijay Vasudevan, Fernanda Viégas, Oriol Vinyals, Pete Warden, Martin Wattenberg, Martin Wicke, Yuan Yu, and Xiaoqiang Zheng. TensorFlow: Large-scale machine learning on heterogeneous systems, 2015. http://tensorflow.org/ 86

Adam Paszke, Sam Gross, Francisco Massa, Adam Lerer, James Bradbury, Gregory Chanan, Trevor Killeen, Zeming Lin, Natalia Gimelshein, Luca Antiga, Alban Desmaison, Andreas Kopf, Edward Yang, Zachary DeVito, Martin Raison, Alykhan Tejani, Sasank Chilamkurthy, Benoit Steiner, Lu Fang, Junjie Bai, and Soumith Chintala. PyTorch: An imperative style, high-performance deep learning library. In *Proc. of the 33rd Conference on Neural Information Processing Systems (NeurIPS)*, pages 8024–8035, 2019. 86

Brian Tanner and Adam White. RL-Glue: Language-independent software for reinforcement-learning experiments. *Journal of Machine Learning Research (JMLR)*, 10:2133–2136, September 2009. 86

Greg Brockman, Vicki Cheung, Ludwig Pettersson, Jonas Schneider, John Schulman, Jie Tang, and Wojciech Zaremba. OpenAI gym. *ArXiv Preprint ArXiv:1606.01540*. 86

Eric Liang, Richard Liaw, Robert Nishihara, Philipp Moritz, Roy Fox, Ken Goldberg, Joseph Gonzalez, Michael Jordan, and Ion Stoica. RLlib: Abstractions for distributed reinforcement learning. In *Proc. of the 35th International Conference on Machine Learning (ICML)*, pages 3053–3062, 2018. 86

Authors' Biographies

FELIPE LENO DA SILVA

Felipe Leno da Silva (Leno) holds a Ph.D. (2019) from the University of São Paulo, Brazil. He is currently a Postdoc Researcher at the Advanced Institute for AI, where he helped organize one of the first Brazilian AI residency programs. He has been actively researching knowledge reuse for multiagent RL since the start of his Ph.D., and is a firm believer that RL will bridge the gap between virtual agents and the physical real world. Leno enjoys serving the AI community in oft-neglected yet important roles. He has been part of the Program Committees of most of the major AI conferences and has organized multiple workshops, such as the Adaptive and Learning Agents (ALA) and the Scaling-Up Reinforcement Learning (SURL) workshop series. Leno is a strong advocate for the inclusion of minorities in the AI community and has been involved in multiple iterations of the Latinx in AI workshop at NeurIPS.

ANNA HELENA REALI COSTA

Anna Helena Reali Costa (Anna Reali) is Full Professor at Universidade de São Paulo (USP), Brazil. She received her Ph.D. at USP, investigated robot vision as a research scientist at the University of Karlsruhe, and was a guest researcher at Carnegie Mellon University, working in the integration of learning, planning, and execution in mobile robot teams. She is the Director of the Data Science Center (C^2D), a partnership between USP and the Itaú-Unibanco bank, and a member of the Center for Artificial Intelligence (C4AI), a partnership between USP, IBM, and FAPESP. Her scientific contributions lie in AI and Machine Learning, in particular RL; her long-term research objective is to create autonomous, ethical, and robust agents that can learn to interact in complex and dynamic environments, aiming at the well-being of human beings.

Printed in the United States
by Baker & Taylor Publisher Services